Neurodynamic Methods for Continuum Robot Control

IEEE Press

445 Hoes Lane

Piscataway, NJ 08854

Neurodynamic Methods for Continuum Robot Control

Ning Tan

School of Computer Science and Engineering, Sun Yat-sen University, Guangzhou, China

Peng Yu

School of Computer Science and Engineering, Sun Yat-sen University, Guangzhou, China

Yunong Zhang

School of Intelligent Systems Engineering, Sun Yat-sen University, Shenzhen, China

IEEE Press
WILEY

Published by John Wiley & Sons, Inc., Hoboken, New Jersey.

Library of Congress Cataloging-in-Publication Data

Names: Tan, Ning, 1984- author | Yu, Peng, 1997- author | Zhang, Yunong author
Title: Neurodynamic methods for continuum robot control / Ning Tan, Peng Yu, Yunong Zhang.
Description: Hoboken, New Jersey: Wiley-IEEE Press, 2026. | Includes bibliographical references and index.
Identifiers: LCCN 2026003590 | ISBN 9781394402731 hardback | ISBN 9781394402755 epdf | ISBN 9781394402748 epub | ISBN 9781394402762 oBook
Subjects: LCSH: Robots—Control systems | Intelligent control systems | Robots—Motion
Classification: LCC TJ211.35. T36 2026
LC record available at https://lccn.loc.gov/2026003590

Cover Design: Wiley
Cover Image: © Ben/stock.adobe.com

Contents

Foreword

Continuum robots, inspired by the extraordinary dexterity and adaptability of biological structures such as octopus tentacles, elephant trunks, and earthworms, have emerged as one of the most promising directions in modern robotics. Over the past decade, the field has witnessed rapid progress in continuum robot design, modeling, perception, and control. Yet despite these advancements, achieving precise, robust, and real-time motion control for such robots remains a formidable challenge, owing primarily to their infinite degrees of freedom, elastic deformation properties, and high sensitivity to environmental disturbances.

Against this backdrop, this book by Ning Tan, Peng Yu, and Yunong Zhang offers a timely and comprehensive contribution to the field. The authors focus on a powerful class of computational models, i.e., neurodynamic approaches, and systematically explore their applications to the motion control of continuum robotic systems. Compared with traditional methods, neurodynamic models such as gradient neurodynamics (GND) and zeroing neurodynamics (ZND) provide intrinsic advantages in online computation, adaptability, and robustness, making them highly suitable for addressing the complexities associated with continuum robots.

This book presents an impressive breadth and depth of research. It begins with foundational concepts and classical neurodynamic methods and then progressively introduces a sequence of innovative frameworks, including model-free controllers, discrete-time implementations, quad neurodynamics, varying-parameter ZND, predefined-time convergent models, fuzzy-enhanced schemes, cerebellum-inspired architectures, and adaptive teleoperation strategies. By integrating theoretical rigor with carefully designed simulations and physical experiments, the authors demonstrate how neurodynamic principles can be leveraged to address long-standing challenges such as Jacobian estimation, inverse kinematics, noise interference, and model uncertainty. Particularly noteworthy is the unified framework proposed for heterogeneous continuum robots, which provides valuable insights for developing generalizable and scalable control algorithms.

Beyond its technical contributions, this book also reflects the authors' long-term dedication to the intersection of continuum robotics and intelligent control. Their work exemplifies how cross-disciplinary thinking, drawing from biology, neural computation, and control theory, can drive innovation in robotics. The results presented here not only enrich the theoretical foundations of neurodynamic control but also hold clear potential for real-world applications, including surgical robotics, industrial human–robot collaboration, and search-and-rescue operations.

As continuum robots continue to mature and move toward broader deployment, the need for efficient, precise, and robust control methods will only grow. I believe this book will serve as a valuable reference for researchers, engineers, and graduate students working in robotics, intelligent systems, and related fields. It offers not merely a summary of existing knowledge, but also a forward-looking perspective that will inspire future research and exploration.

I congratulate the authors on their achievements and am pleased to introduce this book to the robotics community.

Yaonan Wang

Hunan University
Changsha, China
2025

About the Authors

Ning Tan is presently an Associate Professor at Sun Yat-sen University, Guangzhou, China. He received the PhD degree in automation from the Department of Automatic Control and Micro-Mechatronic Systems, Université de Franche-Comté/Franche-Comté Electronics Mechanics Thermal Science and Optics-Sciences and Technologies Institute, Besançon, France, in 2014. From 2014 to 2018, he held a Postdoctoral position and a Research Fellow position at the Singapore University of Technology and Design and the National University of Singapore, Singapore. His research interests include soft robotics, bioinspired design, modular and reconfigurable mechanisms, and micro-nano robotics.

Peng Yu received the BE and ME degrees in software engineering from Sun Yat-sen University in 2019 and 2021, respectively. He is currently a PhD candidate in computer science and technology at the School of Computer Science and Engineering, Sun Yat-sen University. His current research interests include neural networks, robotics, and control theory.

Yunong Zhang received the BS degree in industrial electrical automation from the Huazhong University of Science and Technology, Wuhan, China, in 1996, the MS degree in control theory and control engineering from the South China University of Technology, Guangzhou, China, in 1999, and the PhD degree in mechanical and automation engineering from the Chinese University of Hong Kong, Shatin, Hong Kong, China, in 2003. He is currently a Professor with the School of Intelligent Systems Engineering, Sun Yat-sen University. Before joining Sun Yat-sen University in 2006, he was with the National University of Singapore, Singapore; University of Strathclyde, Glasgow, UK; and National University of Ireland at Maynooth, Galway, Ireland, since 2003. His main research interests include robotics, neural networks, computation, and optimization.

Preface

During the course of robotics research, biology has often provided researchers with inspiration to create novel robotic systems, and continuum robots are a prime example of this. In nature, organisms and structures such as elephant trunks, earthworms, and octopus tentacles exhibit exceptional mobility, manipulation capabilities, and flexibility in complex environments. These biological systems have inspired numerous researchers to replicate such abilities using electromechanical devices. This line of work has significantly contributed to the advancement of continuum robots, making them a focal point of research in recent years. Moreover, continuum robots hold the potential for applications in domains and environments where traditional robotic systems have yet to be effectively deployed. Continuum robots are characterized by their ability to bend continuously, possessing an infinite number of degrees of freedom and elastic structures. Although similar to hyper-redundant robots, continuum robots have distinct differences. Due to their novel structural design, continuum robots have garnered increasing attention within the field of robotics. Their high flexibility and adaptability enable them to serve as both safe and versatile tools in various applications. For instance, in earthquake rescue missions, continuum robots can be deployed to explore unknown, complex environments and search for trapped individuals in tight spaces where it is difficult to determine the presence of survivors. In industrial settings where both robots and human workers operate in the same space, continuum robots can be used for tasks such as grasping and sorting objects. Their soft structure minimizes the risk of injury to human workers, thus significantly enhancing safety. In the medical field, traditional surgical procedures often require large incisions, while continuum robots enable minimally invasive surgery, reducing patient trauma and shortening recovery time. Clearly, continuum robots offer broad potential for application across a wide range of fields.

Motion control of continuum robots involves calculating appropriate actuation signals to ensure the robot follows the desired path and posture. However, several challenges remain in solving this problem. First, the flexible and continuously bending body of a continuum robot presents greater modeling complexity compared to traditional rigid-link robots. This complexity leads to high computational costs during control processes, making it crucial to reduce the computational burden in order to achieve real-time control and broader practical applications. Addressing this challenge is an important yet difficult task. Second, due to the elastic properties of their materials and structures, continuum robots are more susceptible to unpredictable deformations caused by both internal compression and external forces, which can undermine the accuracy of the model. Overcoming the uncertainties in the modeling process is therefore a key issue for continuum robots. Additionally, different types of continuum robots have varied structures and models. If model-based algorithms are used for control, the modeling process must be repeated for each platform, significantly increasing the workload and difficulty of deployment. Therefore, designing a model-independent control algorithm for continuum robots is a vital and challenging task. Finally, the implementation of control systems is often affected by noise or faults, which can compromise the accuracy and reliability of the control system. Designing a control system with high precision and robustness is thus equally important and challenging for continuum robots.

Recurrent neural networks (RNNs) and related neurodynamic approaches have been widely regarded as effective tools for online computation. Among typical neurodynamic models, gradient neurodynamics (GND) has been extensively studied and employed to address various problems, such as solving Lyapunov matrix equations. However, GND does not fully exploit the time derivative information of coefficients, limiting its potential in certain time-invariant tasks. To overcome this limitation, a novel type of neurodynamics known as zeroing neurodynamics (ZND) was proposed. ZND was specifically designed to solve time-varying problems by making full use of time derivative information, leading to enhanced performance and higher accuracy. Due to its adaptability to dynamic environments, ZND has been successfully applied across a range of domains, including time-dependent matrix inversion, time-varying zero-finding problems, and robotic control. Unlike feed-forward neural networks and other learning-based methods, which require large-scale data collection and extensive pretraining, both GND and ZND models can be applied to solve various problems adaptively without the need for pretraining. The design of GND models is based on eliminating scalar-valued norm-based energy functions and is typically formulated in explicit dynamics. In contrast, ZND models aim to eliminate all entries of vector-valued indefinite error functions and are often described by implicit dynamics. Due to their advantages in computational efficiency, precision, and robustness, neurodynamic methods have been widely utilized for motion control in conventional rigid robots, establishing a robust framework in this area. However, the application of neurodynamic methods to the motion control of continuum robots has been relatively limited. A comprehensive and well-established framework for employing neurodynamics in continuum robot control remains underdeveloped, presenting a significant opportunity for further exploration and advancement in this field.

This book provides an in-depth exploration of various neurodynamic approaches for controlling continuum robots. It begins with analytical methods to derive the Jacobian matrix and use neurodynamic approaches for precise kinematic tracking control. Then, we introduce model-free and inverse-free schemes that rely on sensory feedback, including continuous-time and discrete-time algorithms based on GND and ZND. Next, this book advances into sophisticated control frameworks such as quad neurodynamics, varying-parameter ZND (VP-ZND), and predefined-time convergent models, designed to improve tracking performance under uncertain conditions like noise or disturbances. It introduces several hybrid methods, such as the dual fuzzy-enhanced neurodynamics (DFEN), which dynamically adapts control parameters to enhance convergence accuracy and stability. In addition, this book addresses more specialized applications, including cerebellum-inspired network models for visual servoing in surgical robots, and an adaptive teleoperation system that allows human operators to control continuum robots remotely. Overall, the book presents a blend of theoretical innovation and practical applications, offering a comprehensive guide for both researchers and practitioners in continuum robot control.

This book contains 11 chapters that are classified into the following seven parts:

- Part I: Introduction (Chapter 1);
- Part II: Model-based Neurodynamic Methods (Chapter 2);
- Part III: Model-free Neurodynamic Methods (Chapters 3 and 4);
- Part IV: Model- and Inverse-free Zeroing Neurodynamics Methods (Chapter 5);
- Part V: Variants of Zeroing Neurodynamics Methods (Chapters 6–8);
- Part VI: Enhancing Zeroing Neurodynamics Methods with Intelligent Algorithms (Chapters 9 and 10);
- Part VII: Neurodynamics-based Model-free Teleoperation Method (Chapter 11).

Chapter 1 – This chapter begins with foundational modeling techniques of continuum robots, followed by a theoretical introduction to neurodynamics.

Chapter 2 – In this chapter, two types of neurodynamic approaches, namely GND and ZND, are utilized to solve the real-time Jacobian matrix pseudo-inverse problem, thus achieving analytical-Jacobian-based kinematic tracking control of multi-segment continuum robots. Various neurodynamic models are examined, and their performances in terms of tracking accuracy are evaluated under conditions with and without noise disturbances. Simulative validations using a two-segment continuum robot show the feasibility and robustness of these neurodynamic models.

Chapter 3 – This chapter introduces different kinematic controllers and Jacobian estimators for continuum robots, employing ZND and GND, and then presents continuous-time hybrid neurodynamic approaches for continuum robot control. These methods address the trajectory tracking problem for continuum robots, relying solely on user-defined inputs and sensory feedback, allowing for tracking control without the need for model knowledge or internal structure details. Simulations are conducted to validate the effectiveness of the methods.

Chapter 4 – This chapter seeks to develop discrete-time neurodynamic algorithms for the control of continuum robots. To begin with, several discretization methods are introduced, including one-step forward and backward finite difference rules, three-step forward and backward finite difference rules, and a unified m-step forward discretization formula. Subsequently, these methods are utilized to develop discrete GND and discrete ZND algorithms for the control of continuum robots. Finally, simulations and physical experiments are conducted to verify the effectiveness of these algorithms.

Chapter 5 – This chapter introduces a model-free and inverse-free continuous-time quad neurodynamics (CQND) scheme based on ZND. In view of the inherent discrete data processing of computers, the CQND scheme is discretized using various discretization formulas, thereby obtaining a four-instant discrete quad neurodynamics (DQND4) scheme for the control of continuum robots. This chapter also presents the acceleration-level continuous quad neurodynamics (ACQN) control method, a novel strategy designed to enhance adaptability and precision in continuum robots. Unlike conventional velocity-based control frameworks, ACQN operates directly at the acceleration level, ensuring smoother motion trajectories and greater robustness in dynamic environments. To enable practical digital implementation, ACQN is discretized through two complementary approaches: the Verlet method, which provides numerical stability and time-reversibility, and a modified implicit left-and-right 3-step (ILR3S) formula that simplifies derivative computation via the Euler backward approach, yielding the acceleration-level discrete quad neurodynamics (ADQN) method. The effectiveness of these methods is rigorously validated through simulations and physical experiments.

Chapter 6 – This chapter introduces several nonlinear ZND models with enhanced convergence and noise tolerance capability, and a unified dual ZND scheme for kinematic control of continuum robots, thereby enhancing performance under noise. Simulations and experiments validate the feasibility, superior convergence, and robustness of the dual ZND scheme in controlling continuum robots.

Chapter 7 – In this chapter, a model-free scheme based on VP-ZND is presented. The scheme involves two components, one of which solves the inverse kinematics problem based on a VP-ZND model, and the other employs another VP-ZND model to estimate the pseudo-inverse of the Jacobian matrix of continuum robots. In addition, to reduce the dependence on diverse sensory feedback, the recursive least squares method is employed to estimate and iteratively update the pseudo-inverse of the Jacobian matrix based on real-time data. Finally, the feasibility of the control strategies is validated by simulations.

Chapter 8 – This chapter presents a unified control framework for heterogeneous continuum robots based on ZND, with a particular focus on achieving predefined-time convergence and synchronized control. A novel predefined-time convergent ZND (PTC-ZND) model is presented and rigorously analyzed to guarantee error convergence within a user-specified time horizon. This model is further used to construct an inverse kinematics solver and a state estimator, enabling precise and time-bounded motion control. To overcome the lack of synchronization in conventional ZND methods, we introduce a new predefined-time synchronized ZND (PTS-ZND) model that ensures simultaneous convergence of all system states, leading to improved trajectory efficiency and accuracy. Additionally, an online Jacobian learning method based on iterative gradient dynamics is incorporated to enable model-free control. The framework is further extended by integrating prescribed performance control (PPC) theory with ZND, resulting in a behavior-prescribed controller that requires no manual tuning. An adaptive control architecture is then constructed by combining the controller with a ZND-based model estimator. The methods are comprehensively validated across five types of heterogeneous continuum robots, demonstrating strong generalizability, robustness, and practical effectiveness in complex control tasks.

Chapter 9 – In order to effectively enhance the control performance of ZND schemes for continuum robots, a novel DFEN scheme is introduced in this chapter. The scheme utilizes the fuzzy control theory to dynamically adjust the convergence parameters of the ZND scheme, thereby reducing the tracking error of continuum robots. The effectiveness of the fuzzy controller and the DFEN scheme is validated by simulations.

Chapter 10 – This chapter introduces a model-free control scheme designed based on a cerebellum-inspired network model for visual servoing of a surgical rigid-flexible hybrid robotic endoscope. The hybrid robot is composed of a rigid robot manipulator and a flexible endoscope as the end-effector. The scheme combines liquid state machines (LSMs) and ZND, where the LSM is responsible for generating control signals and the ZND is used as the training signal for the LSM. Physical experiments verify that the cerebellum-inspired control scheme is effective to handle the visual servoing task.

Chapter 11 – This chapter presents an adaptive teleoperation system for continuum robots. To this end, a closed-loop control system based on damping ZND (DampZND) is formulated to achieve the kinematic control of continuum robots with unknown kinematics. The control system is constructed using two DampZND models, which are employed to estimate the unknown kinematic model and solve the inverse kinematics problem, respectively. By integrating the DampZND control system with teleoperation technology, this chapter presents an adaptive teleoperation system that allows human operators to easily control continuum slave robots. The efficacy of the teleoperation system is validated through experiments conducted on a cable-driven continuum slave robot.

In summary, this book presents various efficient, precise, and robust neurodynamic methods to address the kinematic control challenges of continuum robotic systems, supported by comprehensive simulations and experimental validations. It is written for researchers and graduate students in academia and industry who are working in the fields of robotics, control theory, and neurodynamics. Beyond its achievements, potential, and perspectives, the book offers a comprehensive view of the combined research of these fields. We sincerely hope that it will inspire and bring enjoyment to readers throughout their studies and research. Without a doubt, this book can still be extended and refined, and any comments or suggestions are warmly welcomed. The authors can be contacted via e-mail at tann5@mail.sysu.edu.cn, yupeng6@mail2.sysu.edu.cn, and zhynong@mail.sysu.edu.cn.

Guangzhou, China
April 2025

Ning Tan, Peng Yu, and Yunong Zhang

Acknowledgments

This book incorporates the results of numerous original research papers conducted by the research group of the authors. Many contributors to these papers have carried out extensive, detailed, and innovative research work. We are deeply grateful to these authors for their high-quality contributions. Throughout the development of this book, we have had the privilege of engaging in in-depth discussions with many collaborators and students, whose input has been invaluable to various aspects and results presented here. We are immensely appreciative of their contributions, which have significantly enhanced the quality of this book.

The continuous support of our research by the National Natural Science Foundation of China (grant numbers 62173352 and 624B2140) and also by the Guangdong Basic and Applied Basic Research Foundation (grant number 2024B1515020104) is gratefully acknowledged.

Besides, we sincerely thank the editors and reviewers for their suggestions. We are very grateful to the nice people at Wiley for their strong support during the preparation and publishing of this book.

Ning Tan, Peng Yu, and Yunong Zhang

Acronyms

ANN	Artificial Neural Network
CR	Continuum Robot
CQND	Continuous-time Quad Neurodynamics
DoF	Degrees of Freedom
DQND	Discrete Quad Neurodynamics
GND	Gradient Neurodynamics
HCR	Heterogeneous Continuum Robot
IK	Inverse Kinematics
LSM	Liquid State Machine
MIS	Minimally Invasive Surgery
RCM	Remote Center of Motion
RMSE	Root-mean-square Error
RNN	Recurrent Neural Network
SNN	Spiking Neural Network
TZD	Taylor–Zhang Discretization
VP-ZND	Varying-parameter Zeroing Neurodynamics
ZND	Zeroing Neurodynamics

Part I

Introduction

1

Introduction to Continuum Robots and Theoretical Foundations

1.1 Introduction

Continuum robots have received increasing attention in recent decades. Their compliance and dexterity make them valuable in various important areas such as medical treatment [1], industry [2], agriculture [3], and aerospace [4]. Fully exploiting their capabilities necessitates the design of effective, efficient, and robust control systems for continuum robots, which remains challenging due to their continuous nature.

Numerous attempts have been made to establish analytical models for continuum robots and develop control strategies based on these models [5–8]. Representative approaches for modeling continuum robots include the piecewise constant-curvature (PCC) model [9], the finite element method [10], the Cosserat rod model [11], and the Euler–Bernoulli beam theory. Based on these established models, control of continuum robots is typically performed using closed-form analytical inverse kinematics solutions [12], solutions computed from the robot Jacobian [13], or solutions obtained through iterative searching. Although model-based control of continuum robots has made significant progress, analytical modeling of continuum robots remains challenging because their configurations are rarely sensed completely and are easily affected by environmental interactions [14], which hinders the application of model-based approaches.

Due to the complexity and difficulty of analytical modeling, numerous data-driven control methods have been proposed to estimate or learn the kinematics, inverse kinematics, or differential kinematics of continuum robots [15–18]. For instance, Thuruthel et al. [19] reported a neural-network-based method to directly learn the inverse kinematics of continuum robots. Kasaei et al. [20] developed a data-efficient neural network to approximate the differential kinematic model and estimate the robot Jacobian for control purposes.

Table 1.1 Notations.

Symbol	Definition
t	Time
$\mathbf{u}$	Actuation state vector of the continuum robot
$\mathbf{p}$	End-effector position of the continuum robot
$\mathbf{p}_{\mathrm{d}}$	Reference position of the continuum robot
$\mathcal{F}$	Matrix of activation functions
$\mathbf{f}$	Kinematic mapping function of the continuum robot
J	Jacobian matrix of the continuum robot
$\|\cdot\|$	2-norm of a vector or Frobenius norm of a matrix
$(\cdot)^{\dagger}$	Pseudoinverse operation
$(\cdot)^{\mathrm{T}}$	Transpose operation

Neurodynamic methods have been employed in the control of continuum robots in recent years due to their advantages such as high efficiency and potential for parallel computation. Existing neurodynamic approaches mainly include gradient neurodynamics (GND) and zeroing neurodynamics (ZND) [21]. The GND method seeks the optimal solution to a control problem by defining a scalar error function and computing its gradient with respect to the task variables. In contrast, the ZND method leverages the time derivative of time-varying variables and specific design formulations to solve control problems. Both methods have been widely applied to various numerical computation problems, such as matrix equations [22], quadratic programming [23], and matrix inversion [24, 25], and have demonstrated promising performance in the control of continuum robots.

This book primarily focuses on how neurodynamic methods can be used to achieve control of continuum robots. To provide readers with a preliminary understanding of the content of this book, this chapter first introduces a basic modeling method for continuum robots based on the PCC model. Then, the theoretical foundations of neurodynamic methods are briefly presented. Some of the notations used in this book are listed in Table 1.1.

1.2 Modeling Continuum Robots Using PCC

For simplicity, this part introduces the kinematic modeling of a multi-segment continuum robot based on the PCC assumption [26], using a series of constant-curvature arcs to approximate the backbone of continuum robots. Based on the PCC assumption, the relationship between actuation space and task space is

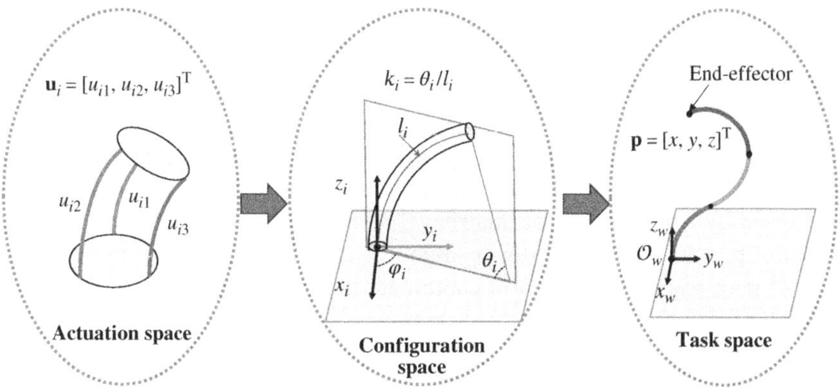

Figure 1.1 Kinematic modeling convention of continuum robots under PCC assumption involves three spaces and two mappings.

decomposed into two parts, as illustrated in Figure 1.1. This involves two mappings: the mapping from actuation space to configuration space, and the mapping from configuration space to task space. The first mapping uses the actuation state vector $\mathbf{u}$ to compute the configuration state (φ, κ, l), where arc parameters φ, κ, and l represent the rotational deflection angle, curvature, and arc length of the arc that denotes a segment of the continuum robot, respectively. In actuation space, $\mathbf{u}_i = [u_{i1}, u_{i2}, u_{i3}]^T$ represents the actuation state of a single segment, where i denotes the ith segment of the multi-segment continuum robot, and u_{ik} describes the state of the kth actuator in the ith segment. In configuration space, a triplet $(\varphi_i, \kappa_i, l_i)$ is defined to represent the arc parameters of the ith segment.

The mapping from actuation space to configuration space is robot-specific, depending on the actuation strategy and generating unique arc parameters in configuration space. This chapter considers a multi-segment continuum robot. The relationship between actuator state and arc parameters of the ith segment is formulated as

$$\varphi_i = \arctan\left(\frac{\sqrt{3}\,(u_{i2} + u_{i3} - 2u_{i1})}{3\,(u_{i2} - u_{i3})}\right), \tag{1.1}$$

$$\kappa_i = \frac{2\sqrt{u_{i1}^2 + u_{i2}^2 + u_{i3}^2 - u_{i1}u_{i2} - u_{i1}u_{i3} - u_{i2}u_{i3}}}{d\,(u_{i1} + u_{i2} + u_{i3})}, \tag{1.2}$$

$$l_i = \frac{u_{i1} + u_{i2} + u_{i3}}{3}, \tag{1.3}$$

where d is the distance between the central backbone of a segment and the actuator.

The mapping from configuration space to task space is robot-independent, which means that it can be applied to all robots that can be approximated as PCC arcs. With the known configuration state, the end-effector position $\mathbf{p}$ can be computed and represented by $\mathbf{p} = [x, y, z]^{\mathrm{T}}$ in task space. The coordinate frames $\mathcal{O}_i$ is attached to the base of the ith segment and the world coordinate frame $\mathcal{O}_w$ is displayed. The transformation between coordinate frames $\mathcal{O}_i$ and $\mathcal{O}_{i+1}$ can be expressed by a transformation matrix $T_i^{i+1} \in \mathrm{SE}(3)$. Therefore, the end-effector position $\mathbf{p}_{\mathrm{e}} \in \mathbb{R}^3$ in the coordinate frame $\mathcal{O}_{n+1}$ is transformed to the position $\mathbf{p} \in \mathbb{R}^3$ with respect to the world coordinate frame $\mathcal{O}_{\mathrm{w}}$ based on the following formulation:

$$\begin{bmatrix} \mathbf{p} \\ 1 \end{bmatrix} = \prod_{i=1}^{n} T_i^{i+1} \begin{bmatrix} \mathbf{p}_{\mathrm{e}} \\ 1 \end{bmatrix}, \tag{1.4}$$

where n is the number of segments and

$$T_i^{i+1} = \begin{bmatrix} C_{\varphi_i}^2 (C_{\kappa_i l_i} - 1) + 1 & S_{\varphi_i} C_{\varphi_i} (C_{\kappa_i l_i} - 1) & C_{\varphi_i} S_{\kappa_i l_i} & \dfrac{C_{\varphi_i} (1 - C_{\kappa_i l_i})}{\kappa_i} \\ S_{\varphi_i} C_{\varphi_i} (C_{\kappa_i l_i} - 1) & C_{\varphi_i}^2 (1 - C_{\kappa_i l_i}) + C_{\kappa_i l_i} & S_{\varphi_i} S_{\kappa_i l_i} & \dfrac{S_{\varphi_i} (1 - C_{\kappa_i l_i})}{\kappa_i} \\ -C_{\varphi_i} S_{\kappa_i l_i} & -S_{\varphi_i} S_{\kappa_i l_i} & C_{\kappa_i l_i} & \dfrac{S_{\kappa_i l_i}}{\kappa_i} \\ 0 & 0 & 0 & 1 \end{bmatrix}, \tag{1.5}$$

with $S_x = \sin(x)$ and $C_x = \cos(x)$.

1.3 Neurodynamic Models

To begin with, the basic design procedures regarding the ZND model and the GND model are introduced.

1.3.1 ZND Model

The foundation of ZND is the error-monitoring function [27], which differs from conventional norm-based scalar-valued error/energy functions. Specifically, the error-monitoring function is indefinite and can be vector- or matrix-valued, as well as real- or complex-valued, depending on the requirements. As an efficient method for solving time-variant problems, ZND involves the following design procedures.

First, an error-monitoring function $\mathbf{e}(t) = \mathbf{x}_\mathrm{d}(t) - \mathbf{x}(t)$, where t denotes time, $\mathbf{x}_\mathrm{d}$ denotes the desired system state and $\mathbf{x}$ denotes the actual system state, must be defined to monitor the process of solving time-variant problems. For brevity, the argument t is dropped hereinafter if no confusion is caused. Then, to ensure the function converges to zero exponentially and globally, its first-order time derivative $\dot{\mathbf{e}}$ is determined using the design formula of ZND as follows:

$$\dot{\mathbf{e}} = \frac{\mathrm{d}\mathbf{e}}{\mathrm{d}t} = -\lambda f(\mathbf{e}), \tag{1.6}$$

where $\lambda \in \mathbb{R}^+$ is a convergence parameter and $f(\cdot)$ is a monotonically increasing odd activation function. Finally, by incorporating the error-monitoring function into the design formula of ZND, the dynamic equation of a ZND model for solving time-variant problems is established.

1.3.2 GND Model

As a typical form of neurodynamics, GND has been extensively studied and applied to solve various problems [28, 29]. For a system where $\mathbf{x}$ denotes the state of the system and $\mathbf{x}_\mathrm{d}$ denotes the desired state, a norm-based scalar energy function $\epsilon = \| \mathbf{x}_\mathrm{d} - \mathbf{x} \|$ related to the GND model is defined based on the specific task, where $\| \cdot \|$ is the 2-norm of a vector or Frobenius norm of a matrix [30]. Then, to evolve along the descent direction (i.e., the negative gradient direction) of the energy function, the GND design formula is as follows:

$$\dot{\mathbf{v}} = -\lambda \frac{\partial \epsilon}{\partial \mathbf{v}}, \tag{1.7}$$

where $\mathbf{v}$ is the independent variable of the system, $\dot{\mathbf{v}}$ denotes the first-order time derivative of $\mathbf{v}$, and $\lambda \in \mathbb{R}^+$ is a convergence parameter. Finally, by expanding the design formula (1.7) with the energy function ϵ, the dynamic equation of a GND model for solving time-variant problems is established.

1.4 Chapter Summary

Taking the PCC assumption as an example, this chapter has introduced a simple modeling methods for continuum robots. Besides, a concise overview of the theoretical foundations of neurodynamic methods, including ZND and GND, is presented to help readers develop a clear understanding of these approaches.

Part II

Model-based Neurodynamic Methods

2

Analytical-Jacobian-based Neurodynamic Control Methods

2.1 Introduction

Kinematic trajectory tracking control of continuum robots is a prevalent challenge in various applications. Numerous researchers have proposed various methods to control continuum robots. For instance, Liu et al. [31] presented an iterative Jacobian-based inverse kinematics method to control the deflection of a catheter equipped with coils on its tip. Cobos-Guzman et al. [32] proposed a novel kinematic approach for controlling the end-effector of a continuum robot in restricted environments. Researchers have made significant efforts to improve the trajectory tracking accuracy of the end-effector [33, 34]. Despite these advancements, the lack of effective kinematic control schemes continues to limit the applications of continuum robots, as most of the aforementioned methods suffer from low accuracy. Therefore, developing an effective control algorithm with higher accuracy for continuum robots remains a significant research challenge.

Model-based control of continuum robots, as one of the main control schemes, must address the inverse kinematics problem by computing the newly required actuator state based on the robot model [35]. Typically, the inverse kinematics problem is addressed at the velocity level by computing the pseudoinverse of time-varying Jacobian matrices of continuum robots. However, this approach may not always meet real-time requirements due to the time-consuming nature of the calculation. Many conventional Jacobian-based control methods for continuum robots overlook this issue and compute the pseudoinverse using traditional methods (e.g., singular value decomposition), which are designed for time-invariant matrices and have higher computational complexity. Motivated by the above problem, this chapter explores methods for obtaining the real-time pseudoinverse of time-varying Jacobian matrices, thereby achieving more accurate and computationally efficient control of continuum robots.

In recent decades, neural networks have garnered significant attention for modeling and controlling continuum robots [36, 37]. Despite various strategies proposed by predecessors for controlling continuum robots, there remains a need to design more accurate control methods. Zeroing neurodynamics (ZND), or zeroing neural network, represents a superior approach that is considered for robot control. ZND-based approaches have been introduced to address issues related to online time-varying problems, robotics, chaos, and more [38]. This means that it has the potential to compute time-varying Jacobian matrix pseudoinverse in real time, so as to achieve the kinematic control. Several robot control algorithms have been proposed based on ZND and its variants [39–42]. However, all of these ZND-based techniques have only been applied to rigid-link robots.

Building on this inspiration, this chapter investigates a series of ZND models to derive the pseudoinverse of the time-varying Jacobian matrix of continuum robots, enabling kinematic tracking control of these robots. It is worth noting that this chapter shares similarities with existing studies. On one hand, this chapter focuses on model-based control of a new generation of robotic systems, namely continuum robots. On the other hand, this chapter is dedicated to analytical-Jacobian-based control of continuum robots, a widely adopted method in the robotics industry. However, the aim of this chapter is to develop a more accurate kinematic control method for continuum robots. Additionally, gradient neurodynamics (GND) models are also adopted to calculate the pseudoinverse of the time-varying Jacobian matrix for comparison. Finally, simulations based on a two-segment continuum robot are performed to verify the feasibility of the control method and the robustness of the ZND models when disturbed by noise.

2.2 Model-based Kinematic Control

This section employs the original ZND (O-ZND), nonlinearly activated ZND (NA-ZND), and nonlinearly activated integration-enhanced ZND (NAIE-ZND) models to calculate the time-varying Jacobian matrix pseudoinverse, thereby achieving the inverse kinematic control of continuum robots. For comparison, the GND model and its nonlinearly activated versions are also presented.

2.2.1 Inverse Kinematics

The kinematics of an n-segment continuum robot can be concisely formulated as

$$\mathbf{f}(\mathbf{u}) = \mathbf{p}, \tag{2.1}$$

where $\mathbf{f}$ denotes the nonlinear kinematic mapping function. To fulfill trajectory tracking tasks, the inverse kinematic solution must be derived based on the desired path $\mathbf{p}_d \in \mathbb{R}^3$. Using the minimum velocity norm scheme [39], the inverse kinematic solution is given by

$$\dot{\mathbf{u}} = J^{\dagger} (\dot{\mathbf{p}}_d + \alpha (\mathbf{p}_d - \mathbf{p})), \quad (2.2)$$

where $\dot{\mathbf{u}}$ and $\dot{\mathbf{p}}_d$ denote the first-order time derivatives of $\mathbf{u}$ and $\mathbf{p}_d$ respectively, $\alpha > 0$ is the convergence rate to adjust the position error feedback, $J = \partial \mathbf{f}(\mathbf{u}) / \partial \mathbf{u} \in \mathbb{R}^{3\times 3n}$ is the Jacobian matrix and $J^{\dagger}$ denotes the right pseudoinverse of the Jacobian matrix. The right pseudoinverse of the Jacobian matrix $J^{\dagger}$ satisfies the equation:

$$J^{\dagger} J J^{\mathrm{T}} = J^{\mathrm{T}}, \quad (2.3)$$

where J^{T} is the transpose of Jacobian matrix. To determine $\dot{\mathbf{u}}$, $J^{\dagger}$ must be computed in real-time during task execution. Neurodynamic approaches, with their parallel computational capability when implemented in hardware circuits, are suitable for such real-time computation. To lay the foundation for future circuit realization, the following neurodynamics models are used to compute the pseudoinverse of the Jacobian matrix, thus obtaining the inverse kinematic solution for continuum robots.

2.2.2 Neurodynamics for Jacobian Inversion

Next, several neurodynamic models are utilized to solve the time-varying Jacobian inversion problem.

2.2.2.1 O-ZND Model

Firstly, for the sake of solving the pseudoinverse of the Jacobian matrix, an error function E is defined as

$$E = XJJ^{\mathrm{T}} - J^{\mathrm{T}}, \quad (2.4)$$

where X denotes the estimation of $J^{\dagger}$. By expanding the evolution law of ZND model, the implicit ZND model for the right pseudoinverse of the Jacobian matrix is attained as follows:

$$\dot{X}JJ^{\mathrm{T}} = -\lambda \mathcal{F}\left(XJJ^{\mathrm{T}} - J^{\mathrm{T}}\right) - X\left(\dot{J}J^{\mathrm{T}} + J\dot{J}^{\mathrm{T}}\right) + \dot{J}^{\mathrm{T}}, \quad (2.5)$$

where $\mathcal{F} \in \mathbb{R}^{3n\times 3}$ is the activation function matrix, $\dot{X}$ and $\dot{J}$ are the first-order time derivatives of X and J, respectively. Different ZND models can be defined by adopting different activation functions. When using the linear activation function $\mathcal{F}_{i,j}(x) = x$, where $\mathcal{F}_{i,j}$ denotes the element in the ith row and jth column of the activation function matrix, the simplest form of ZND model (i.e., O-ZND) is obtained.

The explicit neurodynamic form of O-ZND model and nonlinearly activated (NA) models are as follows:

$$\dot{X} = \dot{X}\left(I - JJ^{\mathrm{T}}\right) - X\left(\dot{J}J^{\mathrm{T}} + J\dot{J}^{\mathrm{T}}\right) - \lambda\mathcal{F}\left(XJJ^{\mathrm{T}} - J^{\mathrm{T}}\right) + \dot{J}^{\mathrm{T}}, \tag{2.6}$$

where $I \in \mathbb{R}^{3\times3}$ is the identity matrix. For the sake of simplicity, we define several matrices as $G = JJ^{\mathrm{T}}$, $Q = I - G$, and $W = J^{\mathrm{T}}$. Then, the above equation is rewritten to the dynamic equation of a neuron as follows:

$$\dot{X}_{i,j} = \sum_{l=1}^{m}\dot{X}_{il}Q_{lj} - \sum_{l=1}^{m}X_{il}\dot{G}_{lj} + \dot{W}_{i,j} - \lambda\mathcal{F}\left(\sum_{l=1}^{m}X_{il}G_{lj} - W_{i,j}\right), \tag{2.7}$$

where $X_{i,j}$, $G_{i,j}$, $Q_{i,j}$, and $W_{i,j}$ are the entries in the ith row and jth column of matrices X, G, Q, and W, respectively.

The generic network structure of the O-ZND model and NA-ZND models is depicted in Figure 2.1.

2.2.2.2 Four Nonlinearly Activated ZND Models

For better convergence and robustness, four types of nonlinear activation functions are adopted in this part, corresponding to four nonlinear ZND models. The first nonlinear activation function is the power-sigmoid activation function (PSAF) [43], which is defined as

$$\mathcal{F}_{i,j}(x) = \begin{cases} x^q, & |x| \geq 1 \\ \dfrac{1+\exp(-\xi)}{1-\exp(-\xi)} \cdot \dfrac{1-\exp(-\xi x)}{1+\exp(-\xi x)}, & |x| < 1 \end{cases}, \tag{2.8}$$

where $q \geq 3$ is any odd integer and $\xi \geq 2$ is a design parameter of the resulting PSAF-ZND model.

Alternatively, the sign-bi-power activation function (SBPF) [44] is introduced to achieve finite-time convergence and defined as follows:

$$\mathcal{F}_{i,j}(x) = \mathrm{sgn}^{\varpi}(x) + \mathrm{sgn}^{\frac{1}{\varpi}}(x), \tag{2.9}$$

where $\varpi \in (0, 1)$ is a design parameter of the resulting SBPF-ZND model and $\mathrm{sgn}^{\varpi}(x)$ is defined as

$$\mathrm{sgn}^{\varpi}(x) = \begin{cases} |x|^{\varpi}, & x > 0 \\ 0, & x = 0 \\ -|x|^{\varpi}, & x < 0 \end{cases}. \tag{2.10}$$

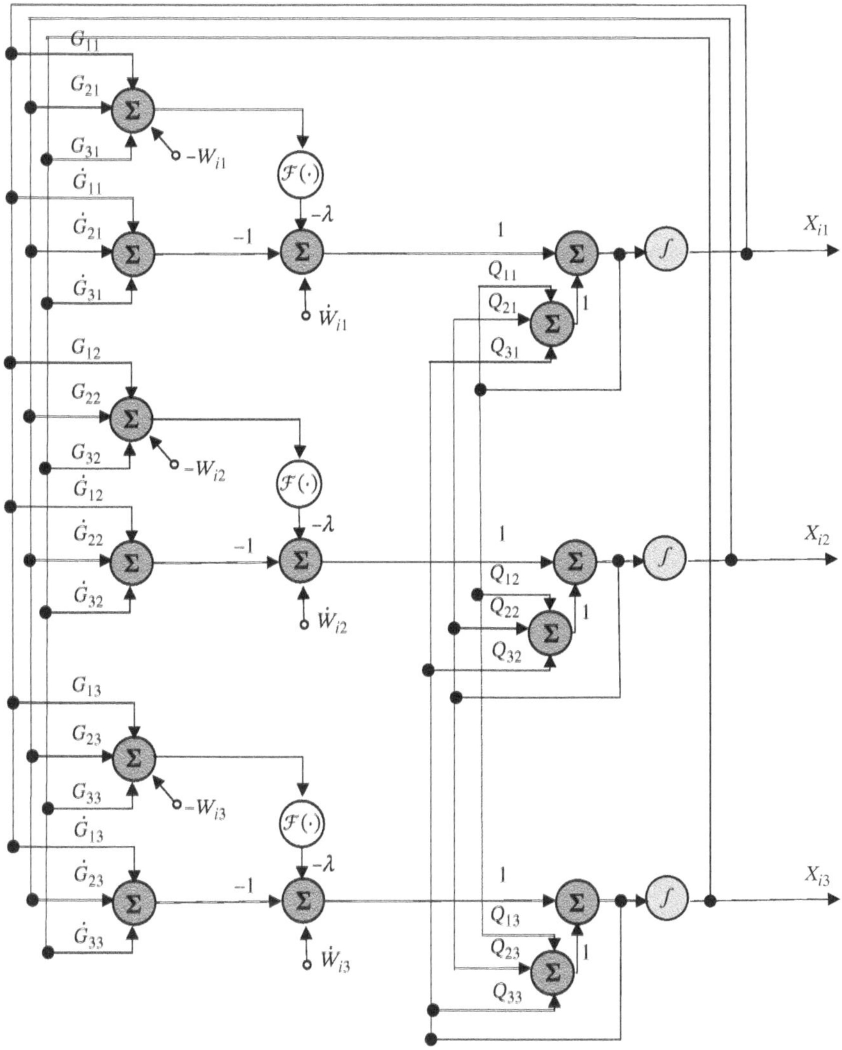

Figure 2.1 Circuit schematics of neuron row i in O-ZND and NA-ZND models.

Based on (2.10), the finite-time activation function (FTAF) [45] is defined as

$$\mathcal{F}_{i,j}(x) = k_1 \text{sgn}^{\varpi}(x) + k_2 x, \tag{2.11}$$

where $k_1 > 0$ and $k_2 > 0$ are parameters of the resulting FTAF-ZND model.

The combined activation function [46] is the combination of sign-bi-power activation function and linear activation function, which is defined as follows:

$$\mathcal{F}_{i,j}(x) = k_1 \mathrm{sbp}(x) + k_2 x, \tag{2.12}$$

where $k_1 > 0$ and $k_2 > 0$ are tunable parameters of the activation function, which will affect the convergence rate of the resulting CAF-ZND model, and

$$\mathrm{sbp}(x) = \mathrm{sgn}^{\varpi}(x) + \mathrm{sgn}^{\frac{1}{\varpi}}(x). \tag{2.13}$$

2.2.2.3 NAIE-ZND Model

Although IE-ZND can handle additional noise, to achieve higher accuracy and finite-time convergence, Tan et al. [47] proposed the following NAIE-ZND model:

$$\dot{E} = -\lambda \mathcal{F}(E) - \mu \int_0^t \mathcal{T}(E(\tau))\mathrm{d}\tau, \tag{2.14}$$

where $\lambda > 0$ and $\mu > 0$ are tunable design parameters of the model. $\mathcal{F}(\cdot) \in \mathbb{R}^{3n\times 3}$ and $\mathcal{T}(\cdot) \in \mathbb{R}^{3n\times 3}$ denote two activation function matrices, both composed of $3n \times 3$ identical monotonically increasing odd activation functions, defined as

$$\begin{cases} \mathcal{F}_{i,j}(x) = \left(k_1 s^{\frac{1}{2}}(t,x)\,\mathrm{sgn}^{\frac{1}{2}}(x) + k_2 s(t,x)\,\mathrm{sgn}(x)\right) \\ \mathcal{T}_{i,j}(x) = \left(k_3 s(t,x)\,\mathrm{sign}(x) + k_4 s^2(t,x)\,\mathrm{sgn}(x)\right) \end{cases}, \tag{2.15}$$

where $k_1, k_2, k_3, k_4 \in \mathbb{R}^+$ are tunable parameters, the function sign$(\cdot)$ denotes the sign function, the first-order time derivative of $s(t, x)$ satisfies $\dot{s}(t, x) = k\mathrm{sign}(x)$ with $k > 0$, and $s(0, x) > 0$. Hence, substituting the same error function into the NAIE-ZND model yields

$$\begin{aligned} \dot{X}JJ^{\mathrm{T}} &= -\lambda \mathcal{F}\left(XJJ^{\mathrm{T}} - J^{\mathrm{T}}\right) + \dot{J}^{\mathrm{T}} - X\left(\dot{J}J^{\mathrm{T}} + J\dot{J}^{\mathrm{T}}\right) \\ &\quad -\mu \int_0^t \mathcal{T}\left(X(\tau)J(\tau)J^{\mathrm{T}}(\tau) - J^{\mathrm{T}}(\tau)\right)\mathrm{d}\tau. \end{aligned} \tag{2.16}$$

Due to the presence of additional noise, the following noise-polluted NAIE-ZND model is investigated

$$\begin{aligned} \dot{X}JJ^{\mathrm{T}} &= -\lambda \mathcal{F}\left(XJJ^{\mathrm{T}} - J^{\mathrm{T}}\right) + \dot{J}^{\mathrm{T}} + N - X\left(\dot{J}J^{\mathrm{T}} + J\dot{J}^{\mathrm{T}}\right) \\ &\quad -\mu \int_0^t \mathcal{T}\left(X(\tau)J(\tau)J^{\mathrm{T}}(\tau) - J^{\mathrm{T}}(\tau)\right)\mathrm{d}\tau, \end{aligned} \tag{2.17}$$

where $N \in \mathbb{R}^{3n\times 3}$ is unknown additional noise.

The network structure of the NAIE-ZND model is depicted in Figure 2.2.

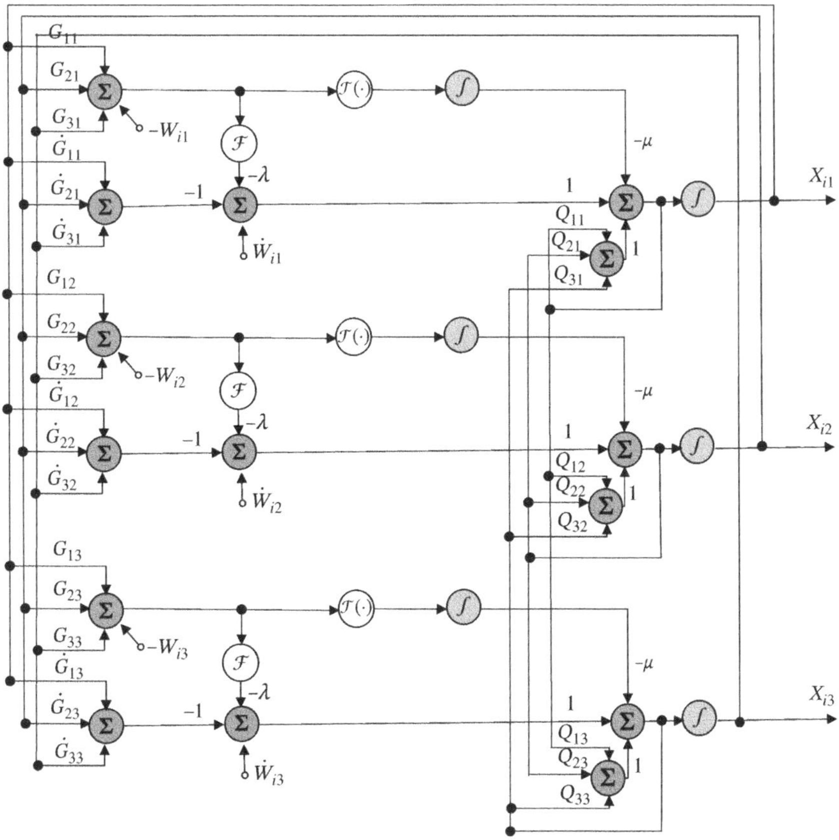

Figure 2.2 Circuit schematics of neuron row i in NAIE-ZND model.

Theorem 2.1[41]: *If the inequality*

$$\left(8\mu k_3 + 9\lambda^2 k_1^2\right)\lambda^2 k_2^2 < 4\mu^2 k_3 k_4 \tag{2.18}$$

holds true, when disturbed by linear noise $N \in \mathbb{R}^{3n\times 3}$ with each element $N_{i,j}$ satisfying $|\dot{N}_{i,j}| \leq c$, where c is a positive constant, the solution generated by the noise-polluted NAIE-ZND model is convergent to the desired solution in finite time.

Proof. By defining

$$\Phi = -\mu \int_0^t T(E(\tau))\mathrm{d}\tau + N, \tag{2.19}$$

the subsystem of NAIE-ZND model is rewritten as

$$\begin{cases} \dot{E}_{i,j} = -\left(\rho_1 S_{i,j}^{\frac{1}{2}} |E_{i,j}|^{\frac{1}{2}} + \rho_2 S_{i,j} |E_{i,j}|\right) \operatorname{sign}(E_{i,j}) + \Phi_{i,j} \\ \dot{\Phi}_{i,j} = -\left(\rho_3 S_{i,j} + \rho_4 S_{i,j}^2 \, |E_{i,j}|\right) \operatorname{sign}(E_{i,j}) + \dot{N}_{i,j} \end{cases}, \tag{2.20}$$

where $\rho_1 = \lambda k_1, \rho_2 = \lambda k_2, \rho_3 = \mu k_3, \rho_4 = \mu k_4$ and $S_{i,j} = s\left(t, E_{i,j}\right)$. Define vector $\mathbf{z}$ and its time derivatives as

$$\mathbf{z} = \begin{bmatrix} z_1 \\ z_2 \\ z_3 \end{bmatrix} = \begin{bmatrix} S_{i,j}^{\frac{1}{2}} |E_{i,j}|^{\frac{1}{2}} \operatorname{sign}(E_{i,j}) \\ S_{i,j} E_{i,j} \\ \Phi_{i,j} \end{bmatrix}, \tag{2.21}$$

$$\dot{\mathbf{z}} = \left(-S_{i,j} V_1 - \frac{S_{i,j}}{|z_1|} V_2 + \frac{\dot{S}_{i,j}}{2S_{i,j}} V_3 + \dot{N}_{i,j} \varrho^{\mathrm{T}}\right) \mathbf{z}, \tag{2.22}$$

where $\varrho = [-\rho_1, \rho_2, 2]^{\mathrm{T}}$ and

$$V_1 = \begin{bmatrix} \upsilon_{11} & 0 & 0 \\ \upsilon_{21} & \rho_2 & -1 \\ \upsilon_{31} & \rho_4 & 0 \end{bmatrix}, \quad V_2 = \begin{bmatrix} \frac{1}{2}\rho_2 & \frac{1}{2}\rho_2 - \upsilon_{11} & -\frac{1}{2} \\ 0 & \rho_1 - \upsilon_{21} & 0 \\ \rho_3 & -\upsilon_{31} & 0 \end{bmatrix}, \quad V_3 = \begin{bmatrix} 1 & 0 & 0 \\ 0 & 2 & 0 \\ 0 & 0 & 0 \end{bmatrix}, \tag{2.23}$$

$$\upsilon_{11} = \frac{\rho_2 \left(3\rho_1^2 \rho_2^2 + 9\rho_1^2 \rho_4 + \rho_2^2 \rho_3 + 4\rho_3 \rho_4\right)}{2\rho_1^2 \rho_4 + 4\rho_2^2 \rho_3 + 16\rho_3 \rho_4}, \tag{2.24}$$

$$\upsilon_{21} = -\frac{\rho_1 \left(3\rho_1^2 \rho_2^2 - \rho_1^2 \rho_4 + 5\rho_2^2 \rho_3 - 8\rho_3 \rho_4\right)}{2(\rho_1^2 \rho_4 + 2\rho_2^2 \rho_3 + 8\rho_3 \rho_4)}, \tag{2.25}$$

$$\upsilon_{31} = \frac{\rho_1 \rho_2 \rho_4 \left(3\rho_1^2 + 7\rho_3\right)}{\rho_1^2 \rho_4 + 2\rho_2^2 \rho_3 + 8\rho_3 \rho_4}. \tag{2.26}$$

Next, a Lyapunov candidate function is defined as

$$w = \mathbf{z}^{\mathrm{T}} H \mathbf{z} = 2\rho_3 z_1^2 + \rho_4 z_2^2 + \frac{1}{2} z_3^2 + \frac{1}{2}(\rho_1 z_1 + \rho_2 z_2 - z_3)^2, \tag{2.27}$$

where

$$H = \frac{1}{2}\begin{bmatrix} \rho_1^2 + 4\rho_3 & \rho_1\rho_2 & -\rho_1 \\ \rho_1\rho_2 & \rho_2^2 + 2\rho_3 & -\rho_2 \\ -\rho_1 & -\rho_2 & 2 \end{bmatrix}. \tag{2.28}$$

The first-order time derivative of the candidate is given by

$$\dot{w} = -S_{i,j}\mathbf{z}^{\mathrm{T}} A_1 \mathbf{z} + \frac{\dot{S}_{i,j}}{2S_{i,j}}\mathbf{z}^{\mathrm{T}} A_3 \mathbf{z} - \frac{S_{i,j}}{|z_1|}\mathbf{z}^{\mathrm{T}} A_2 \mathbf{z} + \dot{N}_{i,j}\boldsymbol{\varrho}^{\mathrm{T}}\mathbf{z}, \tag{2.29}$$

where the matrices are defined as follows:

$$A_1 = \rho_2 \begin{bmatrix} 2\rho_1^2 + \rho_3 & 0 & 0 \\ 0 & \rho_2^2 + \rho_4 & -\rho_2 \\ 0 & -\rho_2 & 1 \end{bmatrix}, \tag{2.30}$$

$$A_2 = \frac{\rho_1}{2}\begin{bmatrix} \rho_1^2 + 2\rho_3 & 0 & -\rho_1 \\ 0 & 5\rho_2^2 + 2\rho_4 & -3\rho_2 \\ -\rho_1 & -3\rho_2 & 1 \end{bmatrix}, \tag{2.31}$$

$$A_3 = \begin{bmatrix} \rho_1^2 + 4\rho_3 & \frac{3\rho_1\rho_2}{2} & -\frac{\rho_1}{2} \\ \frac{3\rho_1\rho_2}{2} & 2\rho_2^2 + 4\rho_4 & -\rho_2 \\ -\frac{\rho_1}{2} & -\rho_2 & 0 \end{bmatrix}. \tag{2.32}$$

According to

$$\lambda_{\min}(H) \parallel \mathbf{z} \parallel_2^2 \leq \mathbf{z}^{\mathrm{T}} H \mathbf{z} \leq \lambda_{\max}(H) \parallel \mathbf{z} \parallel_2^2 \tag{2.33}$$

and

$$|z_1|^2 \leq \| \mathbf{z} \|_2^2 \tag{2.34}$$

where $\lambda_{\min}(H)$ and $\lambda_{\max}(H)$ represent the smallest and largest eigenvalues of matrix H, we have

$$\begin{aligned}
\dot{w} \leq & -\left(S_{i,j} \frac{\lambda_{\min}(A_1)}{\lambda_{\max}(H)} - \frac{\dot{S}_{i,j}\lambda_{\min}(B)}{2S_{i,j}\lambda_{\max}(H)} \right) w \\
& -\left(S_{i,j} \frac{\lambda_{\min}^{\frac{1}{2}}(H)\lambda_{\min}(A_2)}{\lambda_{\max}(H)} - \frac{c\, \|\varrho\|_2}{\lambda_{\min}^{\frac{1}{2}}(H)} \right) w^{\frac{1}{2}}, \\
= & -\sigma w - \gamma w^{\frac{1}{2}}
\end{aligned} \tag{2.35}$$

where

$$B = \begin{bmatrix} \rho_1^2 + 3\rho_1\rho_2/2 + \rho_1/2 + 4\rho_3 & 0 & 0 \\ 0 & 2\rho_2^2 + 3\rho_1\rho_2/2 + \rho_2 + 4\rho_4 & 0 \\ 0 & 0 & \rho_1/2 + \rho_2 \end{bmatrix}. \tag{2.36}$$

If the inequality (2.18) holds true, then $\lambda_{\min}(A_1) > 0, \lambda_{\min}(A_2) > 0, \lambda_{\max}(H) \geq \lambda_{\min}(H) > 0$ and $\lambda_{\max}(B) > 0$. In view of $\dot{S}_{i,j} = k\text{sign}(|E_{i,j}|) \geq 0$ and $S_{i,j}(0) > 0$, $S_{i,j}$ increases with time. Then, σ is positive when $t > t_{g1}$ and γ is positive when $t > t_{g2}$. Therefore, when $t > t_0$ where $t_0 = \max\{t_{g1}, t_{g2}\}$, σ and γ are positive. t_{g1} and t_{g2} are defined as follows:

$$t_{g1} = \begin{cases} 0 & S_{i,j}(0) \geq a \\ \dfrac{a - S_{i,j}(0)}{k} & S_{i,j}(0) < a \end{cases}, \quad t_{g2} = \begin{cases} 0 & S_{i,j}(0) \geq b \\ \dfrac{b - S_{i,j}(0)}{k} & S_{i,j}(0) < b \end{cases}, \tag{2.37}$$

where

$$\begin{aligned}
a &= \frac{k\lambda_{\max}(B)\lambda_{\max}^{\frac{1}{2}}(H)}{2\lambda_{\min}(A_1)\lambda_{\min}^{\frac{1}{2}}(H)}, \\
b &= \frac{c\, \|\varrho\|_2\, \lambda_{\max}(H)}{\lambda_{\min}(H)\lambda_{\min}(A_2)}.
\end{aligned} \tag{2.38}$$

Given that γ increases with time and $\sigma(t_0) \geq 0$, we have

$$\dot{w} < -c_1 w^{\frac{1}{2}}, \; t > t_0, \tag{2.39}$$

where $c_1 = \gamma(t_0) > 0$. Finally, one can get

$$w^{\frac{1}{2}} \begin{cases} < w^{\frac{1}{2}}(t_0) - \dfrac{c_1}{2}(t - t_0), & t < \dfrac{2w^{\frac{1}{2}}(t_0)}{c_1} + t_0 \\ = 0, & t \geq \dfrac{2w^{\frac{1}{2}}(t_0)}{c_1} + t_0 \end{cases}. \tag{2.40}$$

Therefore, w will converge to zero in finite time. Accordingly, the error function $E_{i,j}$ converges to zero within finite time. The proof is thus completed.

2.2.2.4 GND Models

To facilitate comparison with ZND models, the conventional GND model is also investigated in this chapter. The error function is defined as follows:

$$\epsilon = \frac{\| XJJ^{\mathrm{T}} - J^{\mathrm{T}} \|^2}{2}. \tag{2.41}$$

Based on the design formula of GND, X evolves as the following dynamic model

$$\dot{X} = -\lambda \left(XJJ^{\mathrm{T}} - J^{\mathrm{T}} \right) JJ^{\mathrm{T}}, \tag{2.42}$$

where $\lambda > 0$ is the design parameter. Similarly, by using different activation functions, different GND models are obtained, which are formulated as

$$\dot{X} = -\lambda \mathcal{F} \left(XJJ^{\mathrm{T}} - J^{\mathrm{T}} \right) JJ^{\mathrm{T}}, \tag{2.43}$$

where $\mathcal{F}$ can be any of the previously presented activation functions.

2.3 Simulative Verification

In this section, simulations are conducted using the previously discussed inverse kinematic solution (2.2) to complete trajectory tracking tasks with the aid of different neurodynamics-enabled Jacobian inversion methods. To illustrate the feasibility of these methods, simulations are performed on a two-segment continuum robot.

2.3.1 Parameter Setting

The initial actuation state is set as $\mathbf{u}(0) = [0.20, 0.22, 0.18, 0.20, 0.22, 0.18]^{\mathrm{T}}$ m and the initial value of the pseudoinverse of the Jacobian matrix is set as

$$X(0) = \begin{bmatrix} 0.238 & -0.162 & 0.365 \\ 0.411 & 0.081 & 0.385 \\ 0.065 & 0.081 & 0.345 \\ 0.258 & -0.058 & 0.384 \\ 0.133 & 0.029 & 0.117 \\ 0.383 & 0.0289 & 0.651 \end{bmatrix}. \tag{2.44}$$

The convergence rate of the inverse kinematic solver is set as $\alpha = 50$. For the O-ZND and GND models, the convergence parameters are set to $\lambda = 50$ and $\lambda = 10$, respectively. The design parameters of the other four nonlinear function-activated ZND models are set to be the same as those of the O-ZND model. The design parameter λ can be set as large as the hardware permits, as a larger value will result in smaller errors. For the PSAF-ZND model, the parameters are set to $q = 3$ and $\xi = 3$. For the SBPF-ZND, the parameter is set as $\varpi = 0.8$. For the FTAF-ZND model, the parameters are set to $k_1 = k_2 = 1$. For the CAF-ZND model, the parameters are set to $k_1 = k_2 = 1$ and $\varpi = 0.8$. For the IE-ZND model, $\lambda = 50$ and $\mu = 50$, while for the NAIE-ZND model, the parameters $\lambda = 50$, $\mu = 50$, $k = 10$, $k_1 = 1$, $k_2 = 1$, $k_3 = 2$, $k_4 = 5$ are used. The task duration is set to $T_{\mathrm{d}} = 1$ s. These parameters are selected based on the experience from existing works to ensure that these models work well. The reference path is defined as

$$\mathbf{p}_{\mathrm{d}} = \begin{bmatrix} \iota \cos\left(2\pi \sin^2\left(\dfrac{0.5\pi t}{T_{\mathrm{d}}}\right)\right) - \iota \\ \iota \cos\left(\dfrac{\pi}{12}\right) \sin\left(2\pi \sin^2\left(\dfrac{0.5\pi t}{T_{\mathrm{d}}}\right)\right) \\ \iota \sin\left(\dfrac{\pi}{12}\right) \sin\left(2\pi \sin^2\left(\dfrac{0.5\pi t}{T_{\mathrm{d}}}\right)\right) \end{bmatrix} + \mathbf{p}(0), \tag{2.45}$$

where $\iota = 0.17$ m denotes the scale of the trajectory and $\mathbf{p}(0) \in \mathbb{R}^3$ is the initial position of robot end-effector.

2.3.2 Simulation Results

This part elaborates the simulation results of a two-segment continuum robot fulfilling trajectory tracking tasks with the aid of different neurodynamic models.

To begin with, Figure 2.3 shows the specific tracking control results generated by the O-ZND-enabled inverse kinematic solution. Some snapshots of the robot

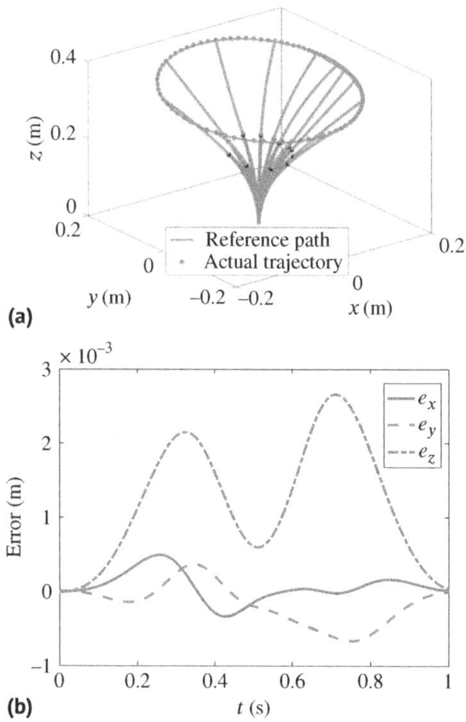

Figure 2.3 Simulation results of two-segment continuum robot tracking circular reference path aided by O-ZND model without noise: (a) snapshots of continuum robot tracking reference path; (b) evolution of tracking errors with respect to different axes.

are depicted in Figure 2.3a and it is observed that the actual end-effector trajectory is consistent with the reference path. At the same time, the evolution of specific tracking errors with respect to different coordinate axes is illustrated in Figure 2.3b. $\mathbf{e}(t) = [e_x, e_y, e_z]^{\mathrm{T}}$ denotes the error between reference path and actual trajectory of the end-effector corresponding to x-, y- and z-axis of the world coordinate system. In short, these results have shown the feasibility of the kinematic tracking scheme for continuum robots.

Next, different neurodynamic models are adopted to fulfill the same tracking task in the noise-free situation for comparison. Simulation results are shown in Figure 2.4, including the tracking errors $\| \mathbf{e}(t) \|$ associated with different ZND and GND models. The root-mean-square error (RMSE) for the O-ZND model is 1.5×10^{-3} m. It is observed that nonlinear activation functions contribute to the

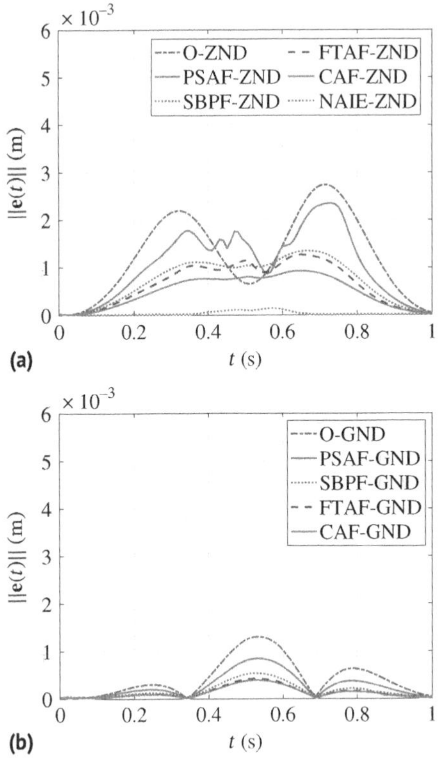

Figure 2.4 Tracking errors $\| \mathbf{e}(t) \|$ associated with different neurodynamic models in noise-free situation: (a) ZND models; (b) GND models.

reduction of tracking error because they can improve the accuracy of Jacobian matrix inversion. Specially, the NAIE-ZND model has achieved the lowest tracking error (4.6×10^{-5} m). Compared to ZND models, GND models have achieved slightly lower errors when adopting the same activation function. For example, the RMSE associated with the O-GND (2.42) is 5.7×10^{-4} m. However, the NAIE-ZND can achieve the best performance due to its special activation functions and the integration operation.

Finally, different neurodynamic models are adopted to fulfill the same tracking task in the noise-polluted situation for comparison. Simulation results are shown in Figure 2.5, including the tracking errors $\| \mathbf{e}(t) \|$ associated with different ZND and GND models. Specifically, the linear increasing noise $N_{i,j} = 10t$ is adopted. It is found that all the ZND and GND models, except for the NAIE-ZND model, are

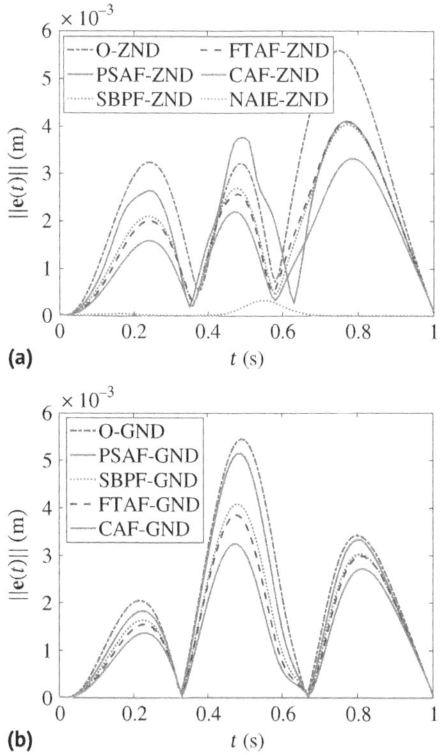

Figure 2.5 Tracking errors || $\mathbf{e}(t)$ || associated with different neurodynamic models in presence of time-varying noise: (a) ZND models; (b) GND models.

Table 2.1 RMSE (unit: mm) associated with different ZND models.

Noise	O-ZND	PSAF-ZND	SBPF-ZND	FTAF-ZND	CAF-ZND	NAIE-ZND
No	1.5	1.3	0.84	0.74	0.58	0.046
Yes	2.9	2.4	2.1	2.1	1.7	0.096

significantly affected by the noise. The tracking errors show an apparent increase in the later stage as a result of the additional noise. However, the NAIE-ZND has been minimally affected, which confirms its superior robustness. Finally, the quantitative comparison in terms of RMSE is summarized in Tables 2.1 and 2.2.

Table 2.2 RMSE (unit: mm) associated with different GND models.

Noise	O-GND	PSAF-GND	SBPF-GND	FTAF-GND	CAF-GND
No	0.57	0.36	0.23	0.18	0.16
Yes	2.6	2.4	2.0	1.9	1.6

2.4 Chapter Summary

In this chapter, various ZND- and GND-based models have been investigated to solve the Jacobian matrix pseudoinverse problem, thus achieving model-based inverse-free kinematic control of continuum robots. Simulation results using a two-segment continuum robot have revealed the feasibility and robustness of the presented control method. Comparisons indicate that, firstly, all the ZND and GND models calculate the pseudoinverse of the Jacobian matrix in real-time, thus achieving kinematic control of continuum robots. Specifically, the NAIE-ZND model achieved the highest accuracy. Secondly, the nonlinear activation functions enhance the convergence rate of neurodynamic models, thereby reducing tracking error. Additionally, the robustness analysis confirmed that all the ZND and GND models are easily affected by noises, except for the NAIE-ZND model.

Part III

Model-free Neurodynamic Methods

3

Continuous-time Hybrid Neurodynamic Control Methods

3.1 Introduction

Tracking control aims at controlling the end-effector of continuum robots to move along the desired path in the workspace. Typically, methods to solve the trajectory tracking problem of robots include kinematic and dynamic methods, the former of which is what this chapter is going to investigate. In the case where the desired path is known, the inverse kinematics method can be used to control the movement of the end-effector. However, there is more than one solution obtained by inverse kinematics due to the infinite degrees of freedom of continuum robots, which makes it difficult to choose an appropriate solution.

Thanks to the efforts of researchers, many methods to overcome or avoid complex inverse kinematics problems have been developed. Cobos-Guzman et al. [32] presented a novel kinematic approach, including inverse kinematic model and forward kinematics for controlling the end-effector of a continuum robot in restricted environments. Lilge and Burgner-Kahrs [48] proposed a kinetostatic modeling framework based on Cosserat rod for parallel continuum robots, thereby solving the forward, inverse, and velocity kinetostatic problems efficiently. It is necessary to know the information of robots' models in advance in the above methods. Besides, the structural complexity of continuum robots makes it difficult to obtain models of continuum robots.

In contrast, model-free methods enable the control of continuum robots without requiring prior model information. Morimoto et al. [49] presented a lightweight ensemble-based reinforcement learning framework for effectively controlling a pneumatically driven soft continuum robot arm with high passive degrees of freedom, demonstrating superior sample efficiency and real-world applicability compared to other reinforcement learning (RL) methods. Melingui et al. [36] designed an adaptive neurodynamic controller that allows the robot to track a desired path in real time with high accuracy. Jin et al. [50] proposed a generalized controller for soft robots, utilizing an estimated Jacobian-based model derived from structural analysis. The model can be simplified based on reasonable assumptions of the robot's structure and updated to balance real-world accuracy with stability. Li et al. [51] developed a model-free method based on an adaptive Kalman filter, used to estimate the robot's Jacobian matrix for path tracking. Liu et al. [52] studied neurodynamics-based tracking control for underactuated systems with unknown parameters and both matched and mismatched disturbances. Significant progress has been made in the control of continuum robots. However, the aforementioned methods may suffer from low accuracy or efficiency, limiting the applications of continuum robots. Therefore, developing a more effective control strategy to improve trajectory tracking accuracy in continuum robots is imperative.

Due to their high performance, recurrent neural networks (RNNs) have been extensively studied and have found wide applications, including machine translation, product recommendations, and robot control [53]. Furthermore, RNNs are capable of solving online time-varying problems [54, 55]. As a novel variant of RNNs, zeroing neurodynamics (ZND) enables the state dimension of the error function to be either multiple or single. This neurodynamic approach ensures exponential convergence in solving time-varying problems by zeroing out each element of the error function [21]. These attributes make ZND highly valuable in practical applications [56].

As a typical RNN, gradient neurodynamics (GND) [57] does not require a training process, as it is based on an energy function and an ordinary differential equation, with constant connection weights. The primary concept behind GND is to minimize an energy function by using the negative gradient of the energy function as the descent direction. During the problem-solving process, the state of GND evolves continuously. Once it reaches a stable state, the output of GND provides the solution to the problem. Recently, GND, along with related neurodynamics, has been extensively studied for solving matrix equations [28, 58], quadratic programming [59], and online time-varying matrix inversion [57]. Furthermore, GND has also been applied to robot control, including the coordination of distributed robots to solve the source-seeking problem [60] and multi-robot competitive coordination [61].

In short, ZND and GND have been applied to the control of a wide range of rigid robots, most of which are model-based. Inspired by these practices, this chapter investigates the application of ZND and GND in the model-free control of continuum robots.

3.2 Model-free Kinematic Control

This section investigates ZND and GND approaches to solve the trajectory tracking problem under the circumstances that the model of the continuum robot is unknown. The kinematic equation of a continuum robot is written as

$$\mathbf{f}(\mathbf{u}(t)) = \mathbf{p}(t), \tag{3.1}$$

where $\mathbf{u}(t) \in \mathbb{R}^m$ denotes the state of m actuators at time t, $\mathbf{p}(t) \in \mathbb{R}^3$ denotes the actual position of the end-effector, and $\mathbf{f}(\cdot) : \mathbb{R}^m \to \mathbb{R}^3$ is a nonlinear mapping. Trajectory tracking is aimed at generating corresponding $\mathbf{u}(t)$ at time t depending on desired path $\mathbf{p}_\mathrm{d}(t) \in \mathbb{R}^3$, such that

$$\mathbf{f}(\mathbf{u}(t)) = \mathbf{p}(t) \to \mathbf{p}_\mathrm{d}(t). \tag{3.2}$$

By differentiating the equation with respect to t, the relationship between the rate of change of actuators and the velocity of the end-effector is obtained as follows:

$$J(t)\dot{\mathbf{u}}(t) = \dot{\mathbf{p}}(t), \tag{3.3}$$

where

$$J(t) = \frac{\partial \mathbf{f}(\mathbf{u}(t))}{\partial \mathbf{u}(t)} \in \mathbb{R}^{3\times m} \tag{3.4}$$

is the Jacobian matrix of the continuum robot and $\dot{\mathbf{p}}_\mathrm{d}(t)$ is the desired velocity of the end-effector at time t. There are two problems that need to be resolved, one of which is the inverse kinematics (IK) problem, and the other is the Jacobian matrix estimation. Both the problems can be solved with neurodynamic models.

An overview of the control scheme composed of ZND and GND approaches is shown in Figure 3.1. The input of the control system is the user-defined path $\mathbf{p}_\mathrm{d}(t)$ and its derivative $\dot{\mathbf{p}}_\mathrm{d}(t)$, and the output is the actual trajectory $\mathbf{p}(t)$ of the robot end-effector. The first step of the scheme is to initialize the actuation signal $\mathbf{u}_0$ and the Jacobian matrix J_0. The controller calculates the actuation signal $\mathbf{u}(t)$ and steers the continuum robot. Meanwhile, the controller calculates the derivatives of the actuation signal (i.e. $\dot{\mathbf{u}}(t)$ and $\ddot{\mathbf{u}}(t)$) as the input of the Jacobian matrix estimation process. The Jacobian matrix $J(t)$ is estimated and transmitted to the controller. The actual position $\mathbf{p}(t)$, actual velocity $\dot{\mathbf{p}}(t)$ and actual acceleration $\ddot{\mathbf{p}}(t)$ of the end-effector are measured by sensors and are treated as feedback information. It is evident that the proposed scheme achieves the tracking control of continuum robots without knowing the model information of robots, which means the scheme is model-free.

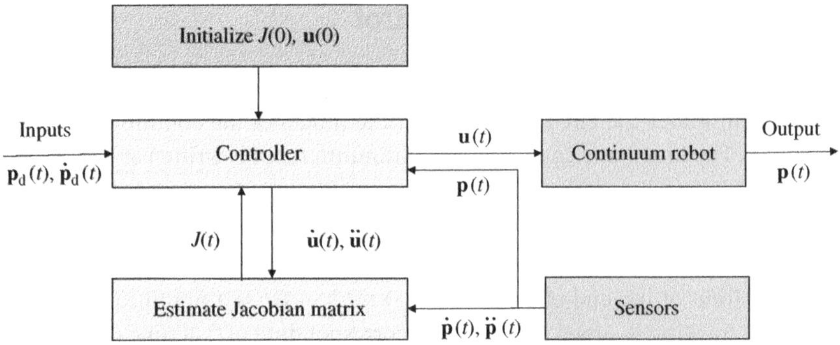

Figure 3.1 Flowchart of hybrid neurodynamic control scheme.

3.2.1 Kinematic Controller

To begin with, two kinematic controllers are developed based on GND and ZND, respectively.

3.2.1.1 GND-based Controller

Firstly, a GND model is used to solve the IK problem. For the solution of the trajectory tracking problem, a scalar-valued norm-based energy function $\epsilon(t)$ is constructed as

$$\epsilon(t) = \frac{1}{2}\|\mathbf{p}_\mathrm{d}(t) - \mathbf{p}(t)\|_2^2, \tag{3.5}$$

where $\|\cdot\|_2$ denotes the 2-norm of a vector. Then, to evolve along the descent direction of the energy function, there is the following GND design formula, which represents the gradient descent direction of $\epsilon(t)$:

$$\begin{aligned}\dot{\mathbf{u}}(t) &= -\gamma\frac{\partial \epsilon(t)}{\partial \mathbf{u}(t)} \\ &= \gamma J^\mathrm{T}(t)\,(\mathbf{p}_\mathrm{d}(t) - \mathbf{p}(t))\end{aligned}, \tag{3.6}$$

where $\gamma > 0$ is the design parameter of the GND model and $J^\mathrm{T}(t)$ is the transpose of the Jacobian matrix.

3.2.1.2 ZND-based Controller

For the purpose of solving the trajectory tracking problem with ZND, a vector-valued error function is defined as follows:

$$\mathbf{e}(t) = \mathbf{p}_\mathrm{d}(t) - \mathbf{p}(t) \in \mathbb{R}^3. \tag{3.7}$$

Next, to make each element of the error function converge to zero, the following formula is chosen as the derivative of the error function via the ZND model:

$$\dot{\mathbf{e}}(t) = -\gamma \mathbf{e}(t), \tag{3.8}$$

where $\dot{\mathbf{e}}(t)$ is the derivative of $\mathbf{e}(t)$ with respect to t and $\gamma > 0$ is a design parameter. With (3.7), the design formula (3.8) is expanded to

$$\dot{\mathbf{p}}_{\mathrm{d}}(t) - J(t)\dot{\mathbf{u}}(t) = -\gamma(\mathbf{p}_{\mathrm{d}}(t) - \mathbf{p}(t)). \tag{3.9}$$

For (3.9), the solution

$$\dot{\mathbf{u}}(t) = J^{\dagger}(t)(\dot{\mathbf{p}}_{\mathrm{d}}(t) + \gamma(\mathbf{p}_{\mathrm{d}}(t) - \mathbf{p}(t))) \tag{3.10}$$

provides a least-square solution for which $\| \mathbf{x}(t) - J(t)\dot{\mathbf{u}}(t) \|$ is the smallest [62] where

$$\mathbf{x}(t) = \dot{\mathbf{p}}_{\mathrm{d}}(t) + \gamma(\mathbf{p}_{\mathrm{d}}(t) - \mathbf{p}(t)). \tag{3.11}$$

For (3.10), $J^{\dagger}(t)$ is the pseudo-inverse of $J(t)$. Output information $\mathbf{p}(t)$ of the robot is measured by sensors. Then it is necessary to determine the value of $J(t)$ to calculate its pseudo-inverse, and thus obtain $\dot{\mathbf{u}}(t)$.

Remark 3.1: It is worth noting that the continuum robot may be in singular configurations in engineering applications, which means the Jacobian matrix is not full rank. In this case, $J^{\mathrm{T}}(t)(\delta I + J(t)J^{\mathrm{T}}(t))$ with δ being a sufficiently small constant and I being an identity matrix, can be employed to approximate the pseudo-inverse $J^{\dagger}(t)$.

3.2.2 Jacobian Matrix Estimation

Evidently, both the above controllers need to determine the Jacobian matrix $J(t)$ of the manipulator in the absence of the kinematic model. Next, this part introduces how to estimate the Jacobian matrix based on neurodynamic approaches.

3.2.2.1 GND-based Jacobian Matrix Estimation

Instead of directly calculating the Jacobian matrix $J(t)$ based on its definition and the models of continuum robots, another GND model is employed for Jacobian matrix estimation. Specifically, another energy function is defined as

$$\varepsilon(t) = \frac{1}{2} \| \dot{\mathbf{p}}(t) - J(t)\dot{\mathbf{u}}(t) \|_2^2. \tag{3.12}$$

Similarly, based on the idea of GND, the following equation is attained:

$$\begin{aligned} \dot{J}(t) &= -\gamma \frac{\partial \varepsilon(t)}{\partial J(t)} \\ &= \gamma\, (\dot{\mathbf{p}}(t) - J(t)\dot{\mathbf{u}}(t))\, \dot{\mathbf{u}}^{\mathrm{T}}(t), \end{aligned} \tag{3.13}$$

where $\dot{J}(t)$ is the derivative of $J(t)$ with respect to t, $\dot{\mathbf{u}}^{\mathrm{T}}(t)$ is the transpose of $\dot{\mathbf{u}}(t)$ and $\gamma > 0$ is another design parameter.

3.2.2.2 ZND-based Jacobian Matrix Estimation

For the purpose of estimating the Jacobian matrix using ZND, a vector-valued error function is defined as follows:

$$\xi(t) = \dot{\mathbf{p}}(t) - J(t)\dot{\mathbf{u}}(t). \tag{3.14}$$

Utilize the ZND approach again and there is

$$\dot{\xi}(t) = -\gamma\xi(t). \tag{3.15}$$

Substituting (3.14) into (3.15) yields

$$\ddot{\mathbf{p}}(t) - \dot{J}(t)\dot{\mathbf{u}}(t) - J(t)\ddot{\mathbf{u}}(t) = -\gamma(\dot{\mathbf{p}}(t) - J(t)\dot{\mathbf{u}}(t)). \tag{3.16}$$

For (3.16), there is the following solution:

$$\dot{J}(t) = (\ddot{\mathbf{p}}(t) - J(t)\ddot{\mathbf{u}}(t) + \gamma(\dot{\mathbf{p}}(t) - J(t)\dot{\mathbf{u}}(t)))\dot{\mathbf{u}}^{\dagger}(t), \tag{3.17}$$

where $\dot{\mathbf{u}}^{\dagger}(t)$ denotes the pseudo-inverse of $\dot{\mathbf{u}}(t)$, $\dot{\mathbf{p}}(t)$ and $\ddot{\mathbf{p}}(t)$ are measured by sensors, and $\ddot{\mathbf{u}}(t)$ is approximated based on the backward Euler method [21]:

$$\ddot{\mathbf{u}}(t) = \frac{(\dot{\mathbf{u}}(t) - \dot{\mathbf{u}}(t - \Delta t))}{\Delta t}. \tag{3.18}$$

Solving the differential equations consisting of a controller and a Jacobian estimator yields the required $\mathbf{u}(t)$ and $J(t)$. Finally, $\mathbf{u}(t)$ is employed to actuate the continuum robot.

3.3 Simulative Verification

In this section, simulations are conducted to validate the feasibility and accuracy of hybrid neurodynamic methods for handling the trajectory tracking problem of continuum robots. By combining different neurodynamic controllers and Jacobian matrix estimators, four different neurodynamic control systems are obtained [63].

3.3.1 Parameter Setting

For simulations, the duration of the entire trajectory tracking task is set as $T_{\mathrm{d}} = 20$ s. The method proposed by Della Santina et al. [35] is adopted to construct a two-segment continuum robot in the simulative environment.

$\mathbf{u}(0) = [3, 0, 0, 3, 0, 0]^{\mathrm{T}}$ mm is the initial state of the actuators. Besides, the design parameters of the neurodynamic models are set as $\gamma = 1$. Theoretically, the larger the value of γ is, the smaller the errors will be. There are two parameters that need to be specified for each segment, i.e., the initial length $L_0 = 100$ mm of each segment and the distance $d = 10$ mm from each actuator to the center axis of robots. To solve differential equations, the initial value of the Jacobian matrix $J(0)$ needs to be determined. By moving each actuator independently by a sufficiently small incremental amount Δu, the position displacement $\Delta \mathbf{p}_i (i = 1, 2, \dots, 6)$ of the end-effector can be measured. Then, the ith column of $J(0)$ is initialized as

$$J_i(0) = \frac{\Delta \mathbf{p}_i}{\Delta u}. \tag{3.19}$$

Without loss of generality, a circular desired path in the three-dimensional space is defined as

$$\mathbf{p}_{\mathrm{d}}(t) = \begin{bmatrix} \kappa \cos\left(2\pi \sin^2\left(\frac{\pi t}{2T_{\mathrm{d}}}\right)\right) - \kappa \\ \kappa \cos\left(\frac{\pi}{6}\right) \sin\left(2\pi \sin^2\left(\frac{\pi t}{2T_{\mathrm{d}}}\right)\right) \\ \kappa \sin\left(\frac{\pi}{6}\right) \sin\left(2\pi \sin^2\left(\frac{\pi t}{2T_{\mathrm{d}}}\right)\right) \end{bmatrix} + \mathbf{p}(0), \tag{3.20}$$

where $\kappa = 50$ mm is the geometry parameter of the desired path $\mathbf{p}_{\mathrm{d}}(t)$ and $\mathbf{p}(0) \in \mathbb{R}^3$ is the absolute position of the path in the task space, i.e., the initial position of the end-effector. The task is to control the end-effector of the continuum robot to move along the above-mentioned desired path.

3.3.2 Simulation Results

Figure 3.2 illustrates the trajectory tracking results of the circular path achieved by the ZND-based controller and ZND-based Jacobian estimator. Figure 3.2a shows the tracking process of the circular path. It is observed that the end-effector strictly follows the desired path. Figure 3.2b illustrates the changes of actuator lengths. These results confirm that the tracking task is fulfilled successfully. The specific tracking results of the four control schemes are shown in Figures 3.3–3.6 respectively, including the comparison between the desired path and actual end-effector trajectory and tracking errors $\mathbf{e} = [e_X, e_Y, e_Z]^{\mathrm{T}}$ with respect to different axes. The quantitative comparisons of different neurodynamic control systems are presented in Table 3.1. Although these hybrid neurodynamic methods can achieve the tracking control of continuum robots, there are two notable results. First, compared to the GND-based controller, the ZND-based controller can achieve lower root-mean-square error (RMSE) and needs less computational cost. Second, compared to the GND-based Jacobian estimator, the ZND-based estimator needs much higher computational cost and more feedback information. Besides, the ZND-based Jacobian estimator can only contribute to the reduction of RMSE when a ZND-based controller is adopted.

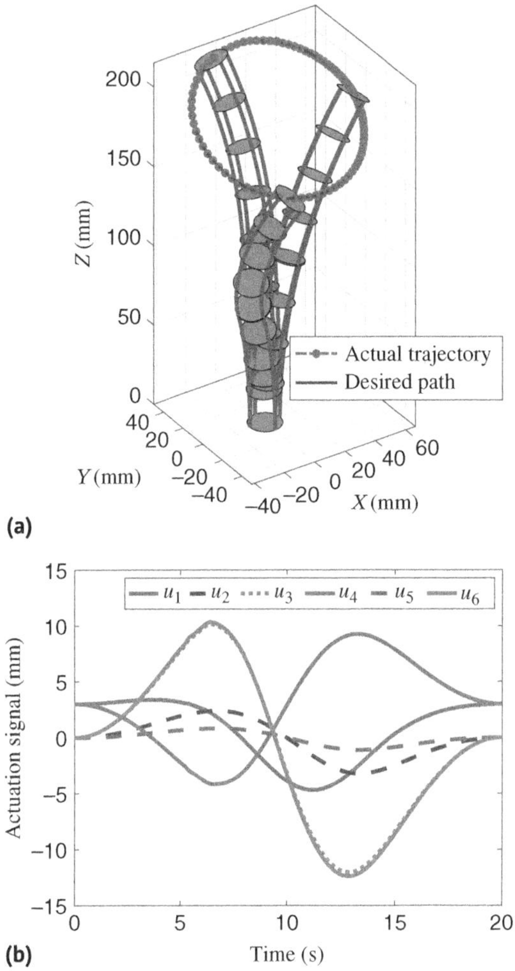

Figure 3.2 Trajectory tracking results of ZND-based controller and ZND-based Jacobian estimator: (a) trajectory tracking process; (b) lengths of actuators.

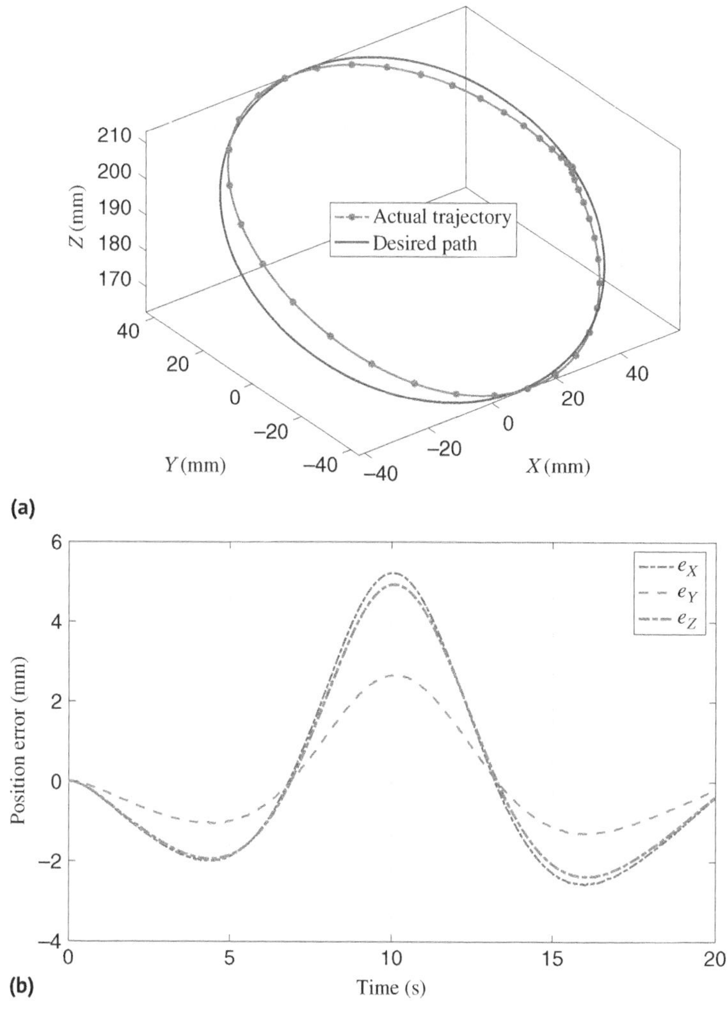

Figure 3.3 Trajectory tracking results of GND-based controller and GND-based Jacobian estimator: (a) trajectory comparisons; (b) tracking errors.

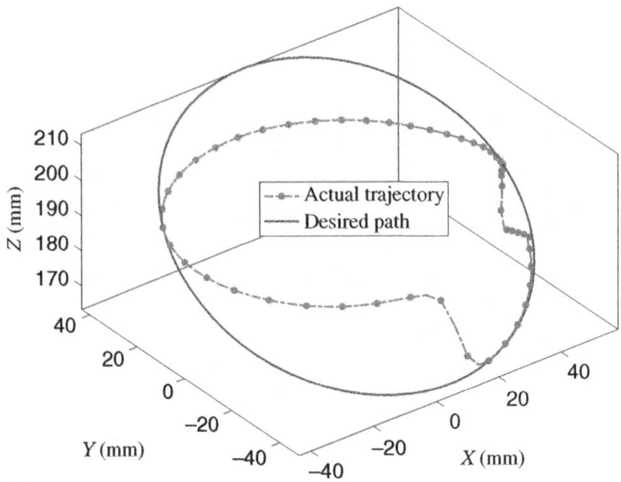

(a)

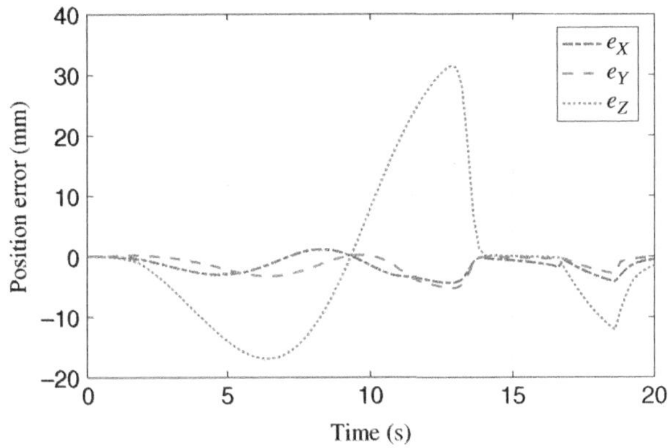

(b)

Figure 3.4 Trajectory tracking results of GND-based controller and ZND-based Jacobian estimator: (a) trajectory comparison; (b) tracking errors.

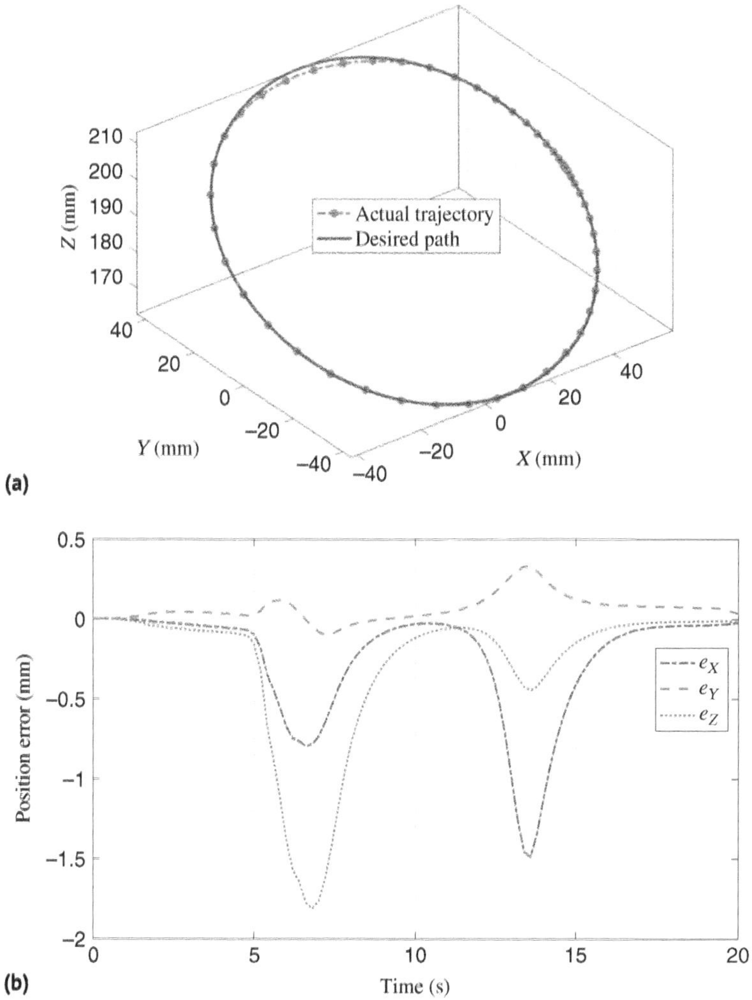

Figure 3.5 Trajectory tracking results of ZND-based controller and GND-based Jacobian estimator: (a) trajectory comparison; (b) tracking errors.

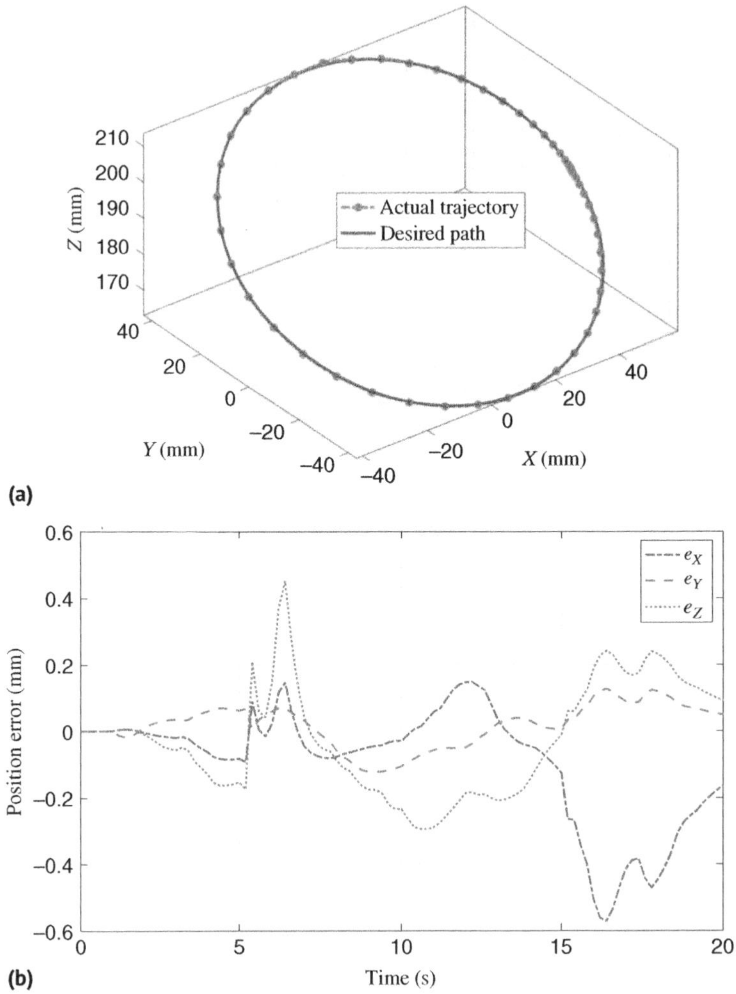

Figure 3.6 Trajectory tracking results of ZND-based controller and ZND-based Jacobian estimator: (a) trajectory comparison; (b) tracking errors.

Table 3.1 Numerical results of hybrid neurodynamic control systems.

Controller	Jacobian estimator	RMSE (mm)	Computational cost (s)
GND	GND	3.52	2.24
GND	ZND	10.69	10.55
ZND	GND	0.70	1.31
ZND	ZND	0.27	8.62

3.4 Chapter Summary

Based on GND and ZND, this chapter introduces different kinematic controllers and Jacobian estimators for continuum robots. By combining these controllers and estimators, four continuous-time model-free neurodynamic schemes are proposed to solve the trajectory tracking problem of continuum robots. In these schemes, the tracking control problem is solved with a neurodynamic model and the Jacobian matrix of continuum robots is estimated with another neurodynamic model, thereby eliminating the requirement for robot kinematics. Finally, simulations based on a two-segment continuum robot are performed to reveal the feasibility and features of these schemes.

4

Discrete-time Neurodynamic Control Methods

4.1 Introduction

In the rapidly evolving field of robotics, continuum robots have emerged as a promising new generation of robotic systems. Inspired by flexible biological structures such as elephant trunks, octopus tentacles, and snakes, these robots offer unique advantages in terms of safety, dexterity, and adaptability [26, 64]. Their ability to navigate complex, unstructured environments and interact safely with humans has sparked interest across various applications, from medical procedures to environmental exploration [9, 65].

However, the control of continuum robots presents significant challenges due to their inherent flexibility and complex dynamics. Traditional control methods, which often rely on rigid-body assumptions, fall short when applied to these highly deformable structures. The continuous deformation of continuum robots, influenced by gravity, external forces, and their weight, necessitates more advanced control strategies [36].

Driven by sensory data, the continuous-time control schemes introduced in the previous chapters are capable of achieving the motion control of continuum robots without knowing the analytical model. However, these continuous-time schemes are not friendly to the deployment in digital devices and the hardware implementation. To solve this issue, the next step is to improve the continuous-time control schemes into discrete-time control algorithms, which have the potential to be implemented in digital circuits with high computational efficiency.

To address these challenges, this chapter focuses on two discrete-time neurodynamic approaches: discrete-time gradient neurodynamics (GND) and discrete-time zeroing neurodynamics (ZND). ZND is an efficient computational framework designed to solve real-time time-varying problems [66]. In the context of continuum robot control, ZND can be used to estimate the robot's Jacobian matrix and solve the inverse kinematics problem without relying on an explicit model of the robot [67]. GND, on the other hand, leverages the concept of energy minimization to solve optimization problems [28]. GND can be applied in continuum robot control to formulate and solve the control problem as an optimization task, allowing for efficient, model-free control [57].

This chapter comprehensively explores discrete-time neurodynamics for continuum robot control. We will discuss discretization techniques for implementing these methods in digital control systems, focusing on maintaining stability and performance in discrete-time domains. The application of ZND-based methods to continuum robot control, including Jacobian estimation and inverse kinematics solving, will be examined in detail. Similarly, we will explore the development of GND-based control strategies, focusing on formulating the control problem as an optimization task. Simulations and experimental results illustrating the effectiveness of ZND and GND-based control on physical continuum robot prototypes will be presented in this chapter. By exploring these advanced neurodynamic approaches, this chapter aims to provide readers with powerful tools for addressing the unique challenges of continuum robot control. The model-free nature of these methods, combined with their ability to operate in discrete-time domains, makes them particularly well-suited for practical implementation in real-world robotic systems.

4.2 Method

This section elaborates the specific theoretical method, including discretization formulas and discrete-time neurodynamic control systems.

4.2.1 Discretization Formulas

Discretization methods are necessary to implement continuous-time neurodynamic methods in digital systems.

To begin with, the one-step forward and backward finite difference rules are presented as follows:

$$\dot{\psi}_k = \frac{\psi_{k+1} - \psi_k}{\tau} + \mathcal{O}(\tau), \tag{4.1a}$$

$$\dot{\psi}_k = \frac{\psi_k - \psi_{k-1}}{\tau} + \mathcal{O}(\tau), \tag{4.1b}$$

where ψ_k denotes the value of a time-varying variable $\psi(t)$ at the kth time step t_k and τ is the sampling interval of time.

Likewise, the three-step forward and backward finite difference rules are expressed as

$$\dot{\psi}_k = \frac{2\psi_{k+1} - 3\psi_k + 2\psi_{k-1} - \psi_{k-2}}{2\tau} + \mathcal{O}(\tau^2), \tag{4.2a}$$

$$\dot{\psi}_k = \frac{3\psi_k - 4\psi_{k-1} + \psi_{k-2}}{2\tau} + \mathcal{O}(\tau^2). \tag{4.2b}$$

Finally, a unified m-step forward discretization formula is presented, which allows for creating discrete-time algorithms with adjustable accuracy. Based on Taylor expansion, for an m-step discrete method, a series of equations can be derived as follows [68]:

$$\psi_{k+1} = \psi_k + \tau\dot{\psi}_k + \frac{(\tau)^2}{2!}\ddot{\psi}_k + \cdots + \frac{(\tau)^m}{m!}\psi^{(m)}(d_1), \tag{4.3a}$$

$$\psi_{k-1} = \psi_k - \tau\dot{\psi}_k + \frac{(\tau)^2}{2!}\ddot{\psi}_k + \cdots + \frac{(\tau)^m}{m!}\psi^{(m)}(d_2), \tag{4.3b}$$

$$\psi_{k-2} = \psi_k - 2\tau\dot{\psi}_k + \frac{(-2\tau)^2}{2!}\ddot{\psi}_k + \cdots + \frac{(-2\tau)^m}{m!}\psi^{(m)}(d_3), \tag{4.3c}$$

$$\psi_{k-m+1} = \psi_k - (m-1)\tau\dot{\psi}_k + \cdots + \frac{(-(m-1)\tau)^m}{m!}\psi^{(m)}(d_m), \tag{4.3d}$$

where $\ddot{\psi}_k$ represents the second-order derivative of ψ_k with respect to t at the time instant t_k. Similarly, $\psi_k^{(m-1)}$ and $\psi_k^{(m)}$ denote the $(m-1)$th-order and mth-order derivatives, respectively. The symbol ! indicates the factorial operator, while d_1, d_2, d_3 and d_m lie in (t_k, t_{k+1}), (t_{k-1}, t_k), (t_{k-2}, t_{k-1}), and (t_{k-m+1}, t_{k-m+2}), respectively.

By eliminating all terms higher than mth-order derivatives of ψ_k in (4.3), a unified m-step forward discretization formula is obtained:

$$\psi_{k+1} = q_1\psi_k + q_2\psi_{k-1} + \cdots + q_m\psi_{k-m+1} + q_{m+1}\tau\dot{\psi}_k + \mathcal{O}(\tau^m), \tag{4.4}$$

where $q_1, q_2, \ldots, q_m$ are parameters that need to be calculated from (4.3). The unified formula provides flexibility in discretizing the continuous-time ZND into discrete-time ZND algorithms with any number of time steps. The choice of m allows users to balance between computational efficiency and accuracy requirements. For instance, $m = 1$ yields a simple one-step method, while higher values of m provide increased accuracy at the cost of more computational complexity.

In the following sections, this chapter presents the application of the above discretization formulas for $m = 1$ and $m = 3$, illustrating how to transform the continuous-time control scheme into discrete-time control algorithms. These examples showcase the versatility of this approach in developing discrete-time neurodynamics-based control strategies.

4.2.2 Robot Kinematics

The kinematic equation of a continuum robot is given by

$$\mathbf{f}(\mathbf{u}(t)) = \mathbf{p}(t), \tag{4.5}$$

where $\mathbf{u}(t)$ denotes the state of actuators at time t, $\mathbf{p}(t) \in \mathbb{R}^3$ denotes the actual position of the end-effector, and $\mathbf{f}(\cdot)$ is a nonlinear mapping. By differentiating the equation with respect to t, the relationship between the change rate of actuators and the velocity of the end-effector is obtained as follows:

$$J(t)\dot{\mathbf{u}}(t) = \dot{\mathbf{p}}(t), \tag{4.6}$$

where $J(t) = \partial\mathbf{f}(\mathbf{u}(t))/\partial\mathbf{u}(t)$ is the Jacobian matrix of the continuum robot.

4.2.3 One-step Discrete-time GND Method

This section introduces a one-step discrete-time GND method. This approach combines the principles of gradient descent with discrete-time implementation, offering an efficient and practical control strategy. Based on the theoretical results in Chapter 3, the continuous-time GND-based controller is formulated as

$$\dot{\mathbf{u}}(t) = -\gamma\frac{\partial\varepsilon(t)}{\partial\mathbf{u}(t)} = \gamma J^{\mathrm{T}}(t)\,(\mathbf{p}_{\mathrm{d}}(t) - \mathbf{p}(t)). \tag{4.7}$$

Meanwhile, the continuous-time GND-based Jacobian estimator is derived as

$$\dot{J}(t) = \gamma\,(\dot{\mathbf{p}}(t) - J(t)\dot{\mathbf{u}}(t))\,\dot{\mathbf{u}}^{\mathrm{T}}(t). \tag{4.8}$$

For discrete-time implementation, the actuation signal update rule becomes

$$\mathbf{u}_{k+1} = \mathbf{u}_k + \tau\dot{\mathbf{u}}_k = \mathbf{u}_k + \alpha_k J_k\left(\mathbf{p}_{\mathrm{d},k} - \mathbf{p}_k\right), \tag{4.9}$$

where $\alpha_k = \gamma\tau$. Notably, the following design parameter α_k is determined dynamically to optimize the control performance [69]:

$$\alpha_k = \frac{\langle \mathbf{e}_k, J_k J_k^{\mathrm{T}}\mathbf{e}_k\rangle}{\langle J_k J_k^{\mathrm{T}}\mathbf{e}_k, J_k J_k^{\mathrm{T}}\mathbf{e}_k\rangle}, \tag{4.10}$$

where $\mathbf{e}_k = \mathbf{p}_{\mathrm{d},k} - \mathbf{p}_k$ and $\langle\cdot,\cdot\rangle$ denotes the inner product of two vectors. Besides, the discrete-time update rule for the estimated Jacobian matrix is

$$J_{k+1} = J_k + \tau\dot{J}_k = J_k + h\,(\dot{\mathbf{p}}_k - J_k\dot{\mathbf{u}}_k)\,\dot{\mathbf{u}}_k^{\mathrm{T}}, \tag{4.11}$$

where $h = \gamma\tau$ is a constant parameter.

By combining (4.9) and (4.11), the one-step discrete-time GND-based control system is obtained.

4.2.4 One-step Discrete-time ZND Method

Based on the theoretical results in Chapter 3, the continuous-time ZND-based controller is formulated as

$$\dot{\mathbf{u}}(t) = J^{\dagger}(t)(\dot{\mathbf{p}}_{\mathrm{d}}(t) + \gamma(\mathbf{p}_{\mathrm{d}}(t) - \mathbf{p}(t))). \tag{4.12}$$

Meanwhile, the continuous-time ZND-based Jacobian estimator is derived as

$$\dot{J}(t) = (\ddot{\mathbf{p}}(t) - J(t)\ddot{\mathbf{u}}(t) + \gamma(\dot{\mathbf{p}}(t) - J(t)\dot{\mathbf{u}}(t)))\dot{\mathbf{u}}^{\dagger}(t). \tag{4.13}$$

Likewise, applying the one-step discretization rules to the continuous-time ZND controller and Jacobian estimator yields the following one-step discrete-time ZND control scheme:

$$\dot{\mathbf{u}}_k = J_k^{\dagger}\left(\dot{\mathbf{p}}_{\mathrm{d},k} + \gamma(\mathbf{p}_{\mathrm{d},k} - \mathbf{p}_k)\right), \tag{4.14a}$$

$$\ddot{\mathbf{u}}_k = \frac{\dot{\mathbf{u}}_k - \dot{\mathbf{u}}_{k-1}}{\tau}, \tag{4.14b}$$

$$\dot{J}_k = (\ddot{\mathbf{p}}_k - J_k\ddot{\mathbf{u}}_k + \gamma(\dot{\mathbf{p}}_k - J_k\dot{\mathbf{u}}_k))\dot{\mathbf{u}}_k^{\dagger}, \tag{4.14c}$$

$$\mathbf{u}_{k+1} = \mathbf{u}_k + \tau\dot{\mathbf{u}}_k, \tag{4.14d}$$

$$J_{k+1} = J_k + \tau\dot{J}_k. \tag{4.14e}$$

Equations (4.14a) and (4.14c) are derived by sampling the continuous-time ZND models at t_k. Equation (4.14b) employs the backward difference approximation (4.1b), while (4.14d) and (4.14e) utilize the forward difference approximation (4.1a).

4.2.5 Three-step Discrete-time ZND Method

The three-step discrete-time ZND control framework is built upon the principles of three-step forward and backward finite difference rules. By applying these formulas in conjunction with previously established equations, the three-step discrete-time ZND control framework is derived for the kth time instant as follows:

$$\dot{\mathbf{u}}_k = J_k^{\dagger}\left(\dot{\mathbf{p}}_{\mathrm{d},k} + \lambda(\mathbf{p}_{\mathrm{d},k} - \mathbf{p}_k)\right), \tag{4.15a}$$

$$\ddot{\mathbf{u}}_k = \frac{\dot{\mathbf{u}}_k - \dot{\mathbf{u}}_{k-1}}{\tau}, \quad (k \le 2), \tag{4.15b}$$

$$\ddot{\mathbf{u}}_k = \frac{3\dot{\mathbf{u}}_k - 4\dot{\mathbf{u}}_{k-1} + \dot{\mathbf{u}}_{k-2}}{2\tau}, \quad (k > 2), \tag{4.15c}$$

$$\dot{J}_k = (\ddot{\mathbf{p}}_k - J_k\ddot{\mathbf{u}}_k + \gamma(\dot{\mathbf{p}}_k - J_k\dot{\mathbf{u}}_k))\dot{\mathbf{u}}_k^{\dagger}, \tag{4.15d}$$

$$\mathbf{u}_{k+1} = \mathbf{u}_k + \tau\dot{\mathbf{u}}_k, \quad (k \leq 2), \tag{4.15e}$$

$$\mathbf{u}_{k+1} = \frac{3}{2}\mathbf{u}_k - \mathbf{u}_{k-1} + \frac{1}{2}\mathbf{u}_{k-2} + \tau\dot{\mathbf{u}}_k, \quad (k > 2), \tag{4.15f}$$

$$J_{k+1} = J_k + \tau\dot{J}_k, \quad (k \leq 2), \tag{4.15g}$$

$$J_{k+1} = \frac{3}{2}J_k - J_{k-1} + \frac{1}{2}J_{k-2} + \tau\dot{J}_k, \quad (k > 2). \tag{4.15h}$$

The equations in the three-step discrete ZND control framework are derived from multiple sources. Equations (4.15a) and (4.15d) result from sampling the dynamic equations of the continuous-time ZND models at discrete-time instant t_k. The three-step backward finite difference formula (4.2b) is the basis for Eq. (4.15c), while Eqs. (4.15f) and (4.15h) are developed using the three-step forward discretization formula (4.2a).

4.3 Verification

In this section, simulations and physical experiments are carried out to validate the above control methods.

4.3.1 Simulations

In this section, simulations are conducted to evaluate the effectiveness of the discrete-time control strategies for continuum robots. Specifically, three control methods, i.e., one-step discrete-time GND, one-step discrete-time ZND, and three-step discrete-time ZND, are applied to a one-segment continuum robot to accomplish trajectory tracking tasks.

The initial actuation state is set as $\mathbf{u}(0) = [0.004, 0, 0]^{\mathrm{T}}$ m, with the tracking task duration set to $T_{\mathrm{d}} = 20$ s and a time step $\tau = 0.01$ s. To initialize the Jacobian matrix $J(0)$, the displacement of the end-effector $\Delta\mathbf{p}_i$ is measured after independently increasing each actuator by a small amount Δc. The ith column of the Jacobian matrix is then initialized as $J_i(0) = \Delta \mathbf{p}_i / \Delta c$. The reference trajectory, formulated as a rhodonea curve, is defined by

$$\mathbf{p}_{\mathrm{d}}(t) = \begin{bmatrix} \xi \cos\left(4\pi\left(\sin^2\left(\frac{\pi t}{2T_{\mathrm{d}}}\right)\right)\right)\cos\left(2\pi\left(\sin^2\left(\frac{\pi t}{2T_{\mathrm{d}}}\right)\right)\right) - \xi \\ \xi \cos\left(\frac{\pi}{6}\right)\cos\left(4\pi\left(\sin^2\left(\frac{\pi t}{2T_{\mathrm{d}}}\right)\right)\right)\sin\left(2\pi\left(\sin^2\left(\frac{\pi t}{2T_{\mathrm{d}}}\right)\right)\right) \\ \xi \sin\left(\frac{\pi}{6}\right)\cos\left(4\pi\left(\sin^2\left(\frac{\pi t}{2T_{\mathrm{d}}}\right)\right)\right)\sin\left(2\pi\left(\sin^2\left(\frac{\pi t}{2T_{\mathrm{d}}}\right)\right)\right) \end{bmatrix} + \mathbf{p}(0), \tag{4.16}$$

where $\xi = 0.04$ m is used to scale the desired path.

To begin with, the three-step discrete-time ZND control scheme is implemented to track the desired path. In this simulation, some configurations of the continuum robot for the tracking task are shown in Figure 4.1a, where the robot's end-effector

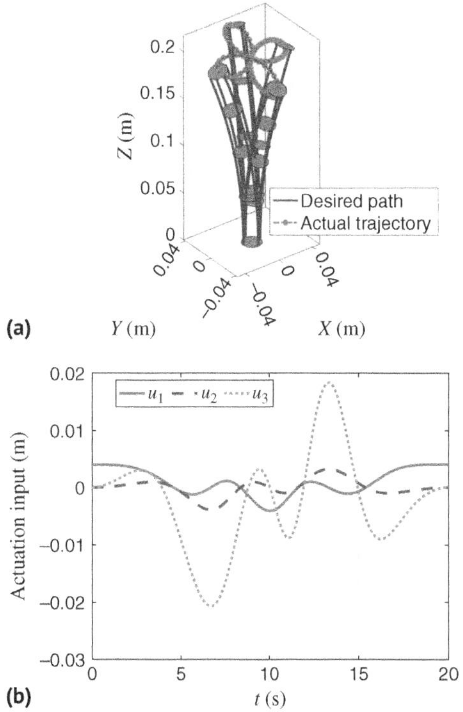

Figure 4.1 Simulation results for continuum robot tracking rhodonea path using three-step discrete-time ZND control scheme: (a) snapshots of robot; (b) actuation input.

accurately follows the desired path. The corresponding actuation input signal is depicted in Figure 4.1b.

Figures 4.2–4.4 illustrate the tracking results for one-step discrete-time GND, one-step discrete ZND, and three-step discrete-time ZND methods, respectively. The tracking error metric, root-mean-square error (RMSE), is used to evaluate tracking accuracy. As shown in Figure 4.2, the one-step GND method achieves an RMSE of 9.33×10^{-4} m, which is the highest error among the three methods. In comparison, the one-step ZND achieves an RMSE of 1.02×10^{-4} m, revealing improved accuracy. The three-step ZND method achieves the lowest tracking error of 2.38×10^{-5} m, indicating the superior control precision.

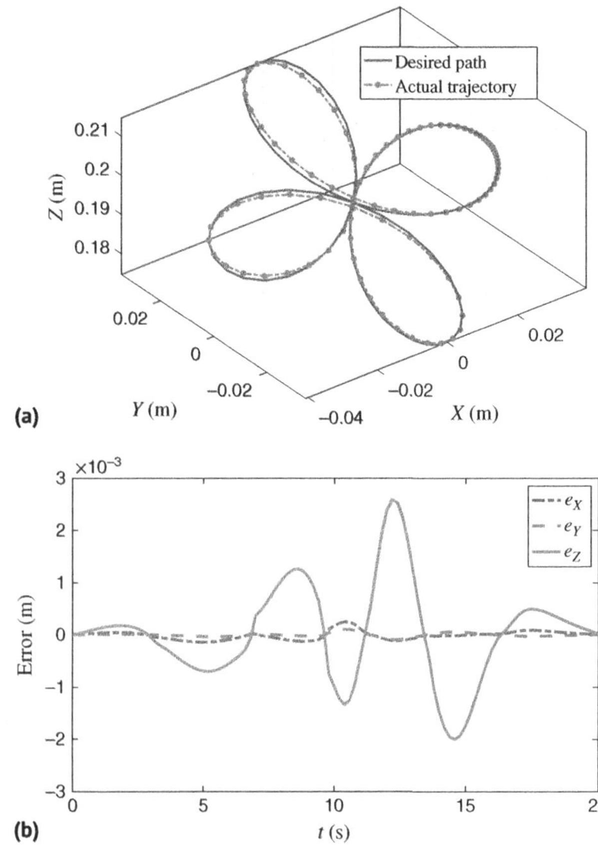

Figure 4.2 Simulation results for continuum robot tracking rhodonea path using one-step discrete-time GND control scheme: (a) trajectory comparison; (b) tracking errors.

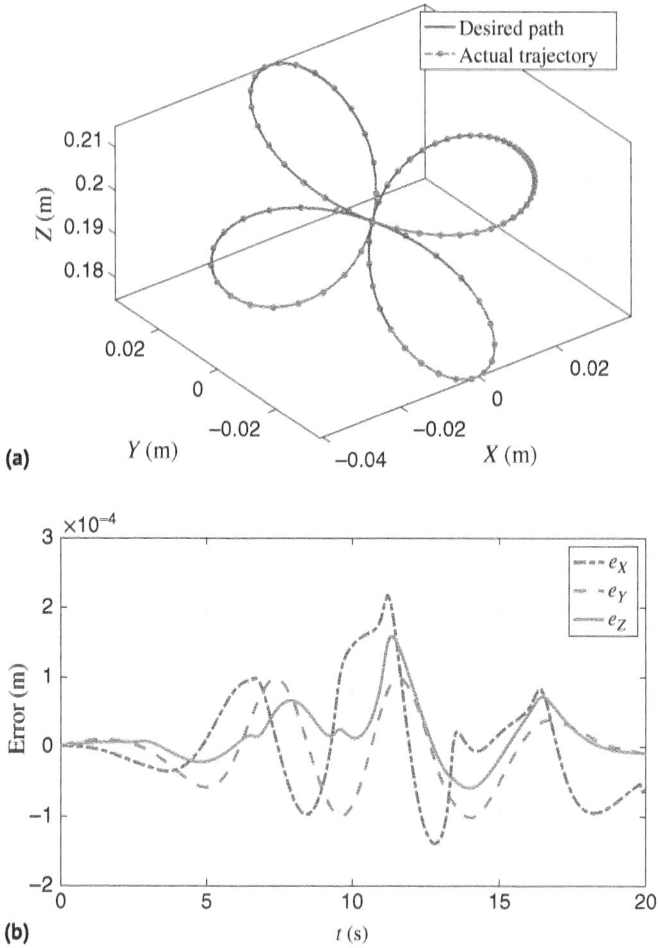

Figure 4.3 Simulation results for continuum robot tracking rhodonea path using one-step discrete-time ZND control scheme: (a) trajectory comparison; (b) tracking errors.

The results highlight the effectiveness of the discrete-time ZND methods in enhancing control accuracy for continuum robots. The three-step ZND method, in particular, shows a significant reduction in tracking error, underscoring the advantage of multi-step discretization in achieving fine-grained control.

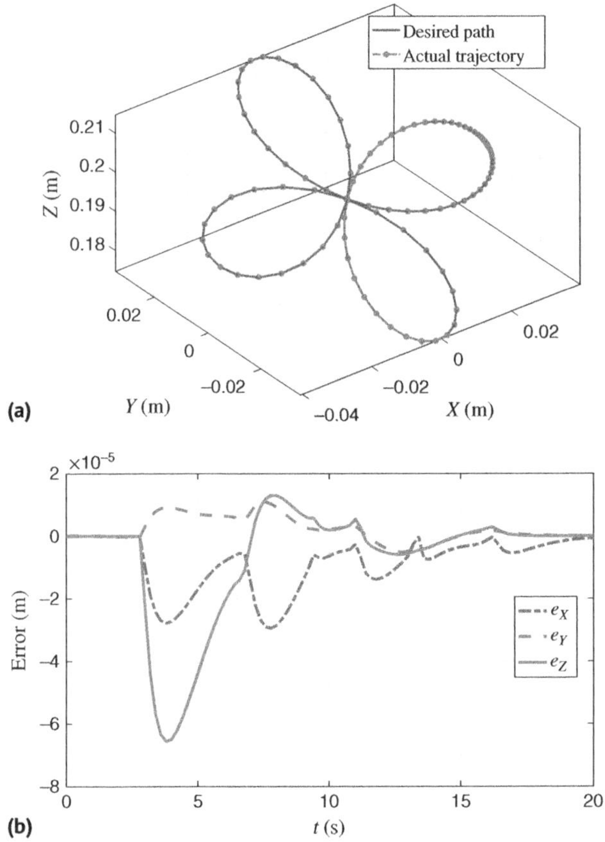

Figure 4.4 Simulation results for continuum robot tracking rhodonea path using three-step discrete-time ZND control scheme: (a) trajectory comparison; (b) tracking errors.

4.3.2 Experiments

To further validate the three-step discrete-time ZND control scheme, experiments are conducted on a physical continuum robot platform. The experimental setup comprises three main components: a computer, a continuum robot, and a visual tracking system. The visual tracking system, MicronTracker H3-60, operates at 16 Hz and provides real-time position feedback of the end-effector to Simulink implemented with the computer.

In physical experiments, the continuum robot is tasked with tracking a pentagram-shaped path, as illustrated in Figure 4.5. The experimental results, including the trajectory tracking result and tracking errors, are presented in Figure 4.6a,b, respectively. The RMSE achieved by the three-step discrete-time

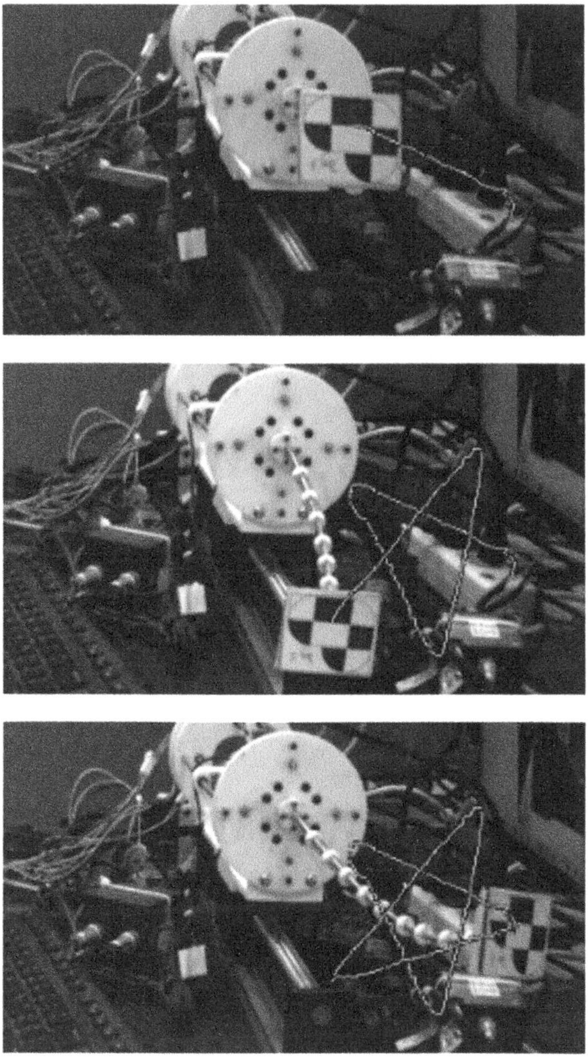

Figure 4.5 Experimental snapshots achieved by three-step discrete-time ZND method for tracking pentagram-shaped path with continuum robot.

ZND method is 1.2×10^{-3} m, showcasing its effectiveness in real-world applications. For comparison, the performance of the one-step GND method and the one-step ZND method is also evaluated. The RMSEs for these methods are 1.1×10^{-2} m (one-step GND) and 3.1×10^{-3} m (one-step ZND), indicating that the three-step ZND-based control method achieves the best accuracy.

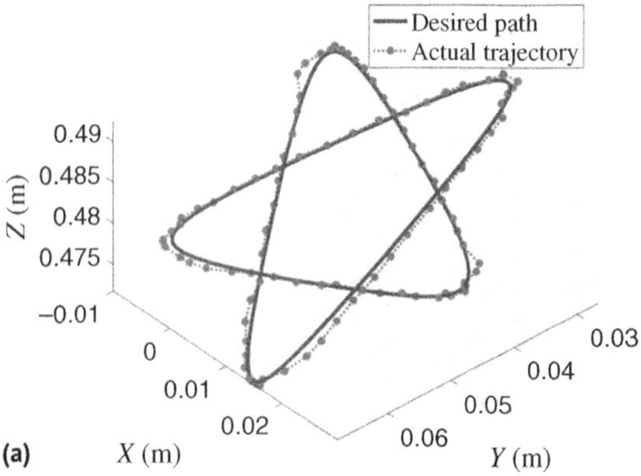

(a)

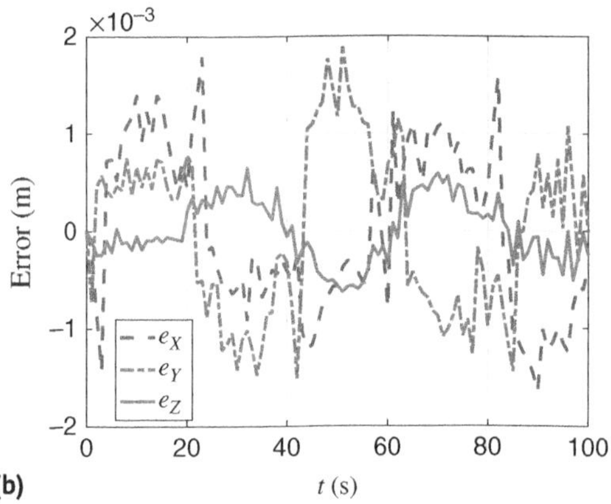

(b)

Figure 4.6 Experimental results achieved by three-step discrete-time ZND method: (a) end-effector trajectories; (b) tracking errors.

4.4 Chapter Summary

In this chapter, several conventional discretization techniques and a unified forward discretization formula have been introduced and derived. Then, these techniques are utilized to improve the continuous-time neurodynamic control schemes to discrete-time neurodynamic control algorithms with an arbitrary number of time steps. Some examples, including one-step discrete-time GND, one-step discrete-time ZND and three-step discrete-time ZND, are provided to illustrate how to develop discrete-time algorithms. Simulation results and experimental results have verified the efficacy of these control algorithms. Specially, different time steps of the discrete control algorithms will lead to different accuracy.

Therefore, users can determine the step number of the discrete-time algorithm according to the requirement on accuracy. Another merit of these algorithms is that they have the potential to be made into digital circuits with high computational efficiency.

Part IV

Model- and Inverse-free Zeroing Neurodynamics Methods

5

Quad Neurodynamic Control Methods

5.1 Introduction

The field of robotics has been revolutionized by the advent of continuum robots, which offer unprecedented flexibility [70] and the ability to navigate complex environments. These remarkable machines can transform various sectors, from minimally invasive surgeries to industrial automation [71]. However, the characteristics that make continuum robots so promising also present significant challenges in their control, especially when their kinematic models still need to be discovered [72].

In recent years, substantial progress has been made in developing control systems for soft continuum manipulators with unknown models [50, 73, 74]. Researchers have explored various approaches. Morimoto et al. [49] developed a reinforcement learning-based controller, and Thuruthel et al. [19] introduced machine learning-based methods for closed-loop kinematic control. These advancements highlight the growing interest in addressing the complexities of controlling continuum robots without predefined models. Nevertheless, the field remains in its infancy, with numerous challenges yet to be fully addressed [75, 76].

In the control schemes presented in previous chapters, the pseudo-inverse of the estimated Jacobian matrix needs to be determined when solving the inverse kinematics problem by zeroing neurodynamics (ZND), which is time-consuming and not friendly to hardware implementation. Therefore, one of the key difficulties in continuum robot control is the need for real-time computation of the Moore–Penrose inverse (i.e., pseudo-inverse) of matrices. This fundamental concept finds applications in various fields, including robotics, signal processing, pattern recognition, and optimization. Numerous algorithms have been proposed for fast matrix Moore–Penrose inversion, such as the Newton's iteration [77], successive matrix square algorithm [78], and Greville recursive method [79].

However, these methods are primarily designed for constant or time-invariant matrices. In practical scenarios, especially when controlling robotic manipulators in real time, we often encounter the need for online solutions involving time-varying matrices. Traditional algorithms struggle to handle such dynamic matrices efficiently. One promising technique to address this challenge is the ZND, which is a Hopfield-type recursive network competent to handle online time-varying problems [80].

Robotic control systems have evolved from basic position-level control to sophisticated hierarchical strategies, each addressing fundamental limitations of prior approaches. Early position control schemes [81] provided reliable joint angle regulation for repetitive tasks but proved inadequate for dynamic environments due to their inability to compensate for disturbances. Velocity-level control [82] emerged as a superior alternative, enabling smoother motion transitions through actuation velocity modulation. However, this approach inherits intrinsic limitations in direct force/torque regulation, particularly problematic for compliant manipulation tasks requiring precise impedance control. This historical progression reveals an ongoing paradigm shift toward acceleration-level frameworks that fundamentally address these force modulation challenges while maintaining motion smoothness.

As extensively documented in the literature [83, 84], conventional velocity-based control methods primarily regulate actuation velocities while fundamentally disregarding their second-order effects on actuation accelerations and torque dynamics. This critical oversight severely constrains their practical implementation in force-sensitive applications, particularly in heavy payload manipulation, where inertial effects dominate system behavior and unstructured environment interactions requiring precise impedance modulation. The inherent limitation stems from the velocity level's inability to explicitly account for the system's dynamic coupling effects, often resulting in either excessive contact forces or unstable interaction dynamics during physical environmental coupling.

To address these challenges, this chapter introduces a continuous-time quad neurodynamics (CQND) model based on ZND. The model comprises four neurodynamic subsystems collaboratively designed to manage complex control tasks of continuum robots when the kinematics model is unknown. Two subsystems are devoted to resolving the inverse kinematics problem and estimating the Jacobian matrix, and the other two focus on calculating the pseudo-inverse of the Jacobian matrix. ZND is particularly effective in defining and systematically reducing an error monitoring function to zero, making it highly suitable for dynamic control scenarios. This efficiency is especially evident in the real-time computation of the Jacobian matrix's pseudo-inverse, leading to enhanced solutions' precision and reduced computational burden. The CQND features an inverse-free approach that enhances computational efficiency and facilitates circuit hardware implementation, which could be difficult in traditional methods because it is difficult to implement matrix inversion with circuits, so the CQND method significantly advances conventional control methods, tackling their inherent challenges more effectively. In view of the inherent discrete data processing of computers, this chapter introduces a four-instant discrete quad neurodynamics (DQND4) model, which aims to align the CQND approach with computer operations, thereby enhancing the practical applicability. The DQND4 model is derived by discretizing the CQND model using Euler's forward and backward formulas, the three-instant backward finite difference formula, and the four-instant Taylor–Zhang discretization (TZD) formula [85]. Finally, a model-free and inverse-free DQND4 method is obtained for the control of continuum robots.

This chapter also presents the acceleration-level continuous-time quad neurodynamics (ACQN) model [86], a novel control framework that extends the principles of CQND to the acceleration domain. Unlike conventional velocity-level methods that often exhibit instability due to indirect acceleration regulation, manifesting as undesirable oscillations and accelerated mechanical wear, the ACQN framework establishes direct acceleration-level control through its innovative ZND architecture. Derived from Hopfield-like dynamics, ZND ensures exponential error convergence while providing unified control for both continuum and rigid robotic systems. The ACQN framework demonstrates superior performance through three key attributes: precise velocity profile regulation that eliminates overshoot and oscillation, adaptive robustness to environmental uncertainties, and significant reduction in mechanical stresses that extends operational lifespan. By directly controlling actuation accelerations, ACQN achieves smoother motion trajectories and enhanced stability compared to traditional approaches, particularly in demanding applications requiring micron-level precision or operation under variable payload conditions. The framework's neurodynamic core enables real-time adaptation to system changes while maintaining strict adherence to joint limits across all kinematic levels – position, velocity, and acceleration – representing a significant advancement in robotic motion control. To enable digital implementation of the ACQN framework, this chapter adopts two specialized discretization methods: the Verlet method [87], valued for its numerical stability and time-reversibility properties, and a modified implicit left-and-right 3-step (ILR3S) formulation that replaces conventional ZND derivatives with an Euler backward approximation [88]. This strategic modification significantly reduces computational overhead while preserving real-time performance, yielding the acceleration-level discrete quad neurodynamics (ADQN) model. These digital implementations retain the inverse-free computation advantage of the original framework, making them particularly suitable for resource-constrained embedded systems. The ADQN architecture exhibits dual-level stability: at the algorithmic level, the continuous control law ensures numerical stability during computation, while at the implementation level, discrete-time robust techniques prevent numerical drift. Rigorous validation through simulations demonstrates the framework's tracking accuracy.

5.2 Preliminary

This section elaborates preliminary knowledge about ZND and robot kinematics.

5.2.1 Zeroing Neurodynamics

ZND, a specialized form of Hopfield-like neural network, is implemented by solving a set of differential equations that govern the system's error dynamics. Unlike conventional numerical solvers that discretize differential equations purely for computation, ZND is fundamentally designed as a dynamic, time-varying problem-solving framework. The core computation in ZND is based on the following differential equation:

$$\dot{\mathbf{e}}(t) = -\gamma \mathbf{e}(t), \tag{5.1}$$

where $\mathbf{e}(t)$ represents the error function between the desired and actual system states, and γ is a positive constant ensuring exponential convergence to zero. The system continuously updates control actions in real-time using a feedback mechanism, where the computed control output adjusts the system state dynamically.

5.2.2 Robot Kinematics

The kinematic equation of a continuum robot is given by

$$\mathbf{f}\,(\mathbf{u}(t)) = \mathbf{p}(t), \tag{5.2}$$

where $\mathbf{u}(t) \in \mathbb{R}^n$ represents the actuation variable of the continuum robot with n actuators, and $\mathbf{p}(t) \in \mathbb{R}^m$ is the actual position of the end-effector in the m-dimensional task space at time t.

In tracking control, the objective is to determine $\mathbf{u}(t)$ at each time t given the desired trajectory $\mathbf{p}_\mathrm{d}(t) \in \mathbb{R}^m$. By differentiating (5.2), a linear relationship at the velocity level is obtained:

$$J(t)\dot{\mathbf{u}}(t) = \dot{\mathbf{p}}(t), \tag{5.3}$$

where $J(t) = \partial \mathbf{f}(\mathbf{u}(t))/\partial \mathbf{u}(t) \in \mathbb{R}^{m\times n}$ is the Jacobian matrix.

5.3 Velocity-level Quad Neurodynamics

This section introduces a velocity-level quad neurodynamics model based on the ZND design formula to compute the actuator state, adaptive Jacobian matrix, and its pseudo-inverse for tracking control of a continuum robot [89].

5.3.1 Continuous-time Quad Neurodynamics

According to the previous chapter, the actuation signal for continuum robots is determined by the following ZND model (termed ZND1):

$$\dot{\mathbf{u}}(t) = J^{\dagger}(t)(\dot{\mathbf{p}}_{\mathrm{d}}(t) + \gamma(\mathbf{p}_{\mathrm{d}}(t) - \mathbf{p}(t))). \tag{5.4}$$

At the same time, the Jacobian matrix of continuum robots is estimated by the following ZND model (termed ZND2):

$$\dot{J}(t) = (\ddot{\mathbf{p}}(t) - J(t)\ddot{\mathbf{u}}(t) + \gamma(\dot{\mathbf{p}}(t) - J(t)\dot{\mathbf{u}}(t)))\frac{\dot{\mathbf{u}}^{\mathrm{T}}(t)}{\dot{\mathbf{u}}^{\mathrm{T}}(t)\dot{\mathbf{u}}(t)}. \tag{5.5}$$

Instead of numerically calculating the pseudo-inverse of the Jacobian in (5.4), the ZND design formula is adopted for the third time. As the right pseudo-inverse $J^{\dagger}(t)$ satisfies $J^{\dagger}(t)J(t)J^{\mathrm{T}}(t) = J^{\mathrm{T}}(t)$, a matrix-valued error function is defined as

$$E = X(t)J(t)J^{\mathrm{T}}(t) - J^{\mathrm{T}}(t), \tag{5.6}$$

where $X(t)$ is the pseudo-inverse of $J(t)$ to be solved. For brevity, the argument t is omitted hereinafter. Applying the ZND design formula again yields

$$\dot{E} = -\gamma E. \tag{5.7}$$

Substituting (5.6) into (5.7) leads to

$$\dot{X}JJ^{\mathrm{T}} + X\left(\dot{J}J^{\mathrm{T}} + J\dot{J}^{\mathrm{T}}\right) - \dot{J}^{\mathrm{T}} = -\gamma\left(XJJ^{\mathrm{T}} - J^{\mathrm{T}}\right). \tag{5.8}$$

The above equation is rewritten to the following explicit form:

$$\dot{X} = \left(-\gamma\left(XJJ^{\mathrm{T}} - J^{\mathrm{T}}\right) - X\left(\dot{J}J^{\mathrm{T}} + J\dot{J}^{\mathrm{T}}\right) + \dot{J}^{\mathrm{T}}\right)\left(JJ^{\mathrm{T}}\right)^{-1}, \tag{5.9}$$

which is termed ZND3.

It is observed from (5.9) that it is still not inverse free. Therefore, we further define

$$W = \left(JJ^{\mathrm{T}}\right)^{-1}. \tag{5.10}$$

To determine the value of W, a matrix-valued error function is defined as

$$\Xi = JJ^{\mathrm{T}} - W^{-1}. \tag{5.11}$$

Utilizing the ZND design formula for the fourth time and substituting (5.11) into it yields

$$\dot{\Xi} = -\gamma\Xi. \tag{5.12}$$

According to the literature, for a time-varying non-singular matrix, we have

$$\dot{W}^{-1} = -W^{-1}\dot{W}W^{-1}. \tag{5.13}$$

By substituting (5.11) into (5.12) and considering (5.13), we obtain the following ZND4 model:

$$\dot{W} = -\gamma\left(WJJ^{\mathrm{T}}W - W\right) - W\left(\dot{J}J^{\mathrm{T}} + J\dot{J}^{\mathrm{T}}\right)W. \tag{5.14}$$

The value of W is obtained by solving the above dynamic equation. Accordingly, the dynamic equation associated with ZND3 is rewritten as follows:

$$\dot{X} = \left(-\gamma\left(XJJ^{\mathrm{T}} - J^{\mathrm{T}}\right) - X\left(\dot{J}J^{\mathrm{T}} + J\dot{J}^{\mathrm{T}}\right) + \dot{J}^{\mathrm{T}}\right)W, \tag{5.15}$$

where X denotes the pseudo-inverse $J^{\dagger}$ of the Jacobian matrix J. Additionally, the dynamic equation associated with ZND1 is rewritten as

$$\dot{\mathbf{u}} = X(\dot{\mathbf{p}}_{\mathrm{d}} + \gamma(\mathbf{p}_{\mathrm{d}} - \mathbf{p})). \tag{5.16}$$

Finally, the CQND method consisting of (5.5), (5.14), (5.15), and (5.16) is formed by utilizing ZND design formulas quadruply. The schematic diagram of the CQND method is illustrated in Figure 5.1.

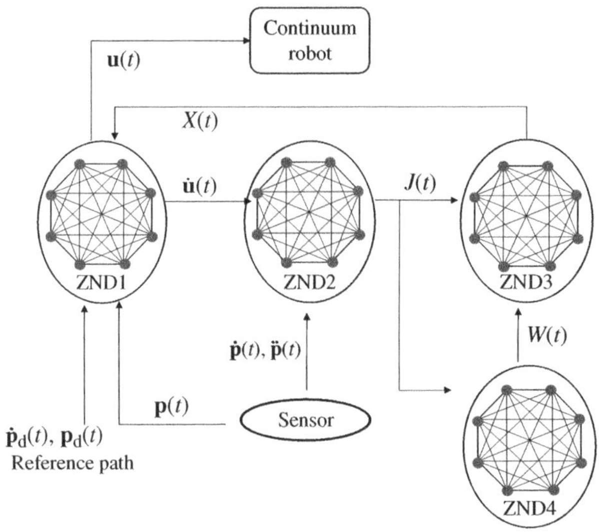

Figure 5.1 Schematic diagram of CQND method.

5.3.2 Discrete Quad Neurodynamics

To facilitate implementation on digital platforms, this chapter presents a discrete version of the CQND method, termed four-instant discrete quad neurodynamics (DQND4) model. This conversion is necessary to simplify computer operations and create a legible program for computing-related variables. Four discretization formulas are employed to derive the DQND4 model, including the four-instant TZD formula, the three-instant backward finite difference discretization formula, the Euler forward formula, and the Euler backward formula. Each of these formulas plays a crucial role in the discretization process, effectively transforming the continuous-time model into a discrete-time version suitable for digital implementation. By utilizing these discretization techniques, the DQND4 model preserves the essential characteristics of the CQND method while adapting it for practical use in digital control systems.

Let r_i denote the value of a time-varying variable r at time instant $t_i = i\tau$, where τ is the time interval. The discretization formulas are presented as follows:

- Four-instant TZD formula:

$$\dot{r}_i = \frac{2r_{i+1} - 3r_i + 2r_{i-1} - r_{i-2}}{2\tau} + O(\tau^2). \tag{5.17}$$

- Three-instant backward finite difference formula:

$$\dot{r}_i = \frac{3r_i - 4r_{i-1} + r_{i-2}}{2\tau} + O(\tau^2). \tag{5.18}$$

- Euler forward formula:

$$\dot{r}_i = \frac{r_{i+1} - r_i}{\tau} + O(\tau). \tag{5.19}$$

- Euler backward formula:

$$\dot{r}_i = \frac{r_i - r_{i-1}}{\tau} + O(\tau), \tag{5.20}$$

where $O(\tau)$ represents the truncation error.

To calculate the variables (i.e., $\mathbf{u}_i$, J_i, X_i, and W_i) for the first two steps, the following Euler forward formulas are adopted:

$$\mathbf{u}_{i+1} = \mathbf{u}_i + \tau\dot{\mathbf{u}}_i, \tag{5.21a}$$

$$J_{i+1} = J_i + \tau\dot{J}_i, \tag{5.21b}$$

$$X_{i+1} = X_i + \tau\dot{X}_i, \tag{5.21c}$$

$$W_{i+1} = W_i + \tau\dot{W}_i. \tag{5.21d}$$

Algorithm 1: Description of DQND4 algorithm.

Initialize $\mathbf{u}_0, J_0, X_0, W_0$ and $\ddot{\mathbf{u}}_0$
Set $T, h, \tau, \mathbf{p}_\mathrm{d}$
for $i = 0 : \frac{T}{\tau} - 1$ **do**
  Calculate $\dot{\mathbf{u}}_i$ via (5.22a)
  Calculate $\dot{J}_i$ via (5.22a)
  Calculate $\dot{X}_i$ via (5.22a)
  Calculate $\dot{W}_i$ via (5.22a)
  if $i \leq 2$ **then**
    Calculate $\mathbf{u}_{i+1}$ via (5.21a)
    Calculate J_{i+1} via (5.21b)
    Calculate X_{i+1} via (5.21c)
    Calculate W_{i+1} via (5.21d)
    Calculate $\ddot{\mathbf{u}}_i$ via (5.23)
  else
    Calculate $\mathbf{u}_{i+1}$ via (5.24a)
    Calculate J_{i+1} via (5.24b)
    Calculate X_{i+1} via (5.24c)
    Calculate W_{i+1} via (5.24d)
    Calculate $\ddot{\mathbf{u}}_i$ via (5.24e)
  end if
end for

The values of $\dot{\mathbf{u}}_i$, $\dot{J}_i$, $\dot{X}_i$, and $\dot{W}_i$ are calculated by sampling (5.16), (5.5), (5.15), and (5.14) at t_i, respectively:

$$\dot{\mathbf{u}}_i = X_i \left(\dot{\mathbf{p}}_{\mathrm{d},i} + \frac{h}{\tau}(\mathbf{p}_{\mathrm{d},i} - \mathbf{p}_i) \right), \tag{5.22a}$$

$$\dot{J}_i = \left(\ddot{\mathbf{p}}_i - J_i \ddot{\mathbf{u}}_i + \frac{h}{\tau}(\dot{\mathbf{p}}_i - J_i \dot{\mathbf{u}}_i) \right) \frac{\dot{\mathbf{u}}_i^\mathrm{T}}{\dot{\mathbf{u}}_i^\mathrm{T} \dot{\mathbf{u}}_i}, \tag{5.22b}$$

$$\dot{X}_i = \left(-\frac{h}{\tau} \left(X_i J_i J_i^\mathrm{T} - J_i^\mathrm{T} \right) - X_i \left(\dot{J}_i J_i^\mathrm{T} + J_i \dot{J}_i^\mathrm{T} \right) + \dot{J}_i^\mathrm{T} \right) W_i, \tag{5.22c}$$

$$\dot{W}_i = -\frac{h}{\tau} \left(W_i J_i J_i^\mathrm{T} W_i - W_i \right) - W_i \left(\dot{J}_i J_i^\mathrm{T} + J_i \dot{J}_i^\mathrm{T} \right) W_i, \tag{5.22d}$$

where $h = \gamma\tau$. The Euler backward formula (5.20) is used to calculate $\ddot{\mathbf{u}}_i$ for the first two steps as follows:

$$\ddot{\mathbf{u}}_i = \frac{\dot{\mathbf{u}}_i - \dot{\mathbf{u}}_{i-1}}{\tau}. \tag{5.23}$$

For subsequent steps during the task, to achieve higher accuracy, the discrete-time states of the above variables are calculated using the four-instant TZD formula (5.17) and the three-instant backward finite difference formula (5.18):

$$\mathbf{u}_{i+1} = \frac{3}{2}\mathbf{u}_i - \mathbf{u}_{i-1} + \frac{1}{2}\mathbf{u}_{i-2} + \tau\dot{\mathbf{u}}_i, \tag{5.24a}$$

$$J_{i+1} = \frac{3}{2}J_i - J_{i-1} + \frac{1}{2}J_{i-2} + \tau\dot{J}_i, \tag{5.24b}$$

$$X_{i+1} = \frac{3}{2}X_i - X_{i-1} + \frac{1}{2}X_{i-2} + \tau\dot{X}_i, \tag{5.24c}$$

$$W_{i+1} = \frac{3}{2}W_i - W_{i-1} + \frac{1}{2}W_{i-2} + \tau\dot{W}_i, \tag{5.24d}$$

$$\ddot{\mathbf{u}}_i = \frac{3\dot{\mathbf{u}}_i - 4\dot{\mathbf{u}}_{i-1} + \dot{\mathbf{u}}_{i-2}}{2\tau}. \tag{5.24e}$$

Notably, the DQND4 method is entirely inverse free, making it computationally efficient and suitable for implementation on digital platforms.

5.4 Acceleration-level Quad Neurodynamics

This section elaborates the theoretical foundation for the acceleration-level quad neurodynamics.

5.4.1 ACQN Method

For end-effector trajectory tracking, we define a vector-valued error function representing the difference between desired and actual end-effector velocities:

$$\mathbf{e}(t) = \dot{\mathbf{p}}_{\mathrm{d}}(t) - \dot{\mathbf{p}}(t). \tag{5.25}$$

By employing the integration-enhanced ZND formulation, we ensure asymptotic convergence of all error components to zero:

$$\dot{\mathbf{e}}(t) = -\gamma\left(\mathbf{e}(t) + \int_0^t \mathbf{e}\,\mathrm{d}\tau\right), \tag{5.26}$$

where $\dot{\mathbf{e}}(t)$ is the derivative of $\mathbf{e}(t)$, and $\gamma \in \mathbb{R}^+$ is a parameter controlling the convergence rate.

Substituting (5.25) into (5.26), the following equation is obtained:

$$\ddot{\mathbf{p}}_{\mathrm{d}}(t) - \dot{J}(t)\dot{\mathbf{u}}(t) - J(t)\ddot{\mathbf{u}}(t) = -\gamma\left((\dot{\mathbf{p}}_{\mathrm{d}}(t) - \dot{\mathbf{p}}(t)) + (\mathbf{p}_{\mathrm{d}}(t) - \mathbf{p}(t))\right). \tag{5.27}$$

Rewriting the above equation gives

$$J(t)\ddot{\mathbf{u}}(t) = \ddot{\mathbf{p}}_{\mathrm{d}}(t) - \dot{J}(t)\dot{\mathbf{u}}(t) + \gamma\left((\dot{\mathbf{p}}_{\mathrm{d}}(t) - \dot{\mathbf{p}}(t)) + (\mathbf{p}_{\mathrm{d}}(t) - \mathbf{p}(t))\right) \tag{5.28}$$

and

$$\ddot{\mathbf{u}}(t) = J^{\dagger}(t)\left(\ddot{\mathbf{p}}_{\mathrm{d}}(t) - \dot{J}(t)\dot{\mathbf{u}}(t) + \gamma\left((\dot{\mathbf{p}}_{\mathrm{d}}(t) - \dot{\mathbf{p}}(t)) + (\mathbf{p}_{\mathrm{d}}(t) - \mathbf{p}(t))\right)\right), \tag{5.29}$$

where $\ddot{\mathbf{u}}(t)$ represents the actuation acceleration at time t, $J^{\dagger}(t)$ denotes the pseudo-inverse of the Jacobian matrix $J(t)$, and $\dot{\mathbf{u}}(t)$ is the actuation velocity. Additionally, $\dot{\mathbf{p}}_{\mathrm{d}}(t)$ and $\dot{\mathbf{p}}(t)$ denote the desired and actual end-effector velocities, respectively. The neurodynamic model formulated in (5.29) is denoted as AZND1.

To compute the pseudo-inverse of the Jacobian matrix, one must first obtain the Jacobian matrix itself. However, since the robot's model is assumed to be unknown in this scenario, the Jacobian matrix cannot be directly calculated, as the robot's forward kinematic mapping is unavailable.

To estimate the unknown Jacobian matrix, the ZND design formula is applied again. The following vector-valued error function is defined:

$$\epsilon(t) = \dot{\mathbf{p}}(t) - J(t)\dot{\mathbf{u}}(t). \tag{5.30}$$

Applying the ZND design formula $\dot{\epsilon}(t) = -\gamma\epsilon(t)$ to (5.30) yields:

$$\dot{J}(t) = \left(\ddot{\mathbf{p}}(t) - J(t)\ddot{\mathbf{u}}(t) + \gamma\left(\dot{\mathbf{p}}(t) - J(t)\dot{\mathbf{u}}(t)\right)\right)\frac{\dot{\mathbf{u}}^{\mathrm{T}}(t)}{\dot{\mathbf{u}}^{\mathrm{T}}(t)\dot{\mathbf{u}}(t)}, \tag{5.31}$$

where $\dot{\mathbf{p}}(t)$ and $\ddot{\mathbf{p}}(t)$ denote the velocity and acceleration of the end-effector, respectively. By solving the dynamic equation (5.31), denoted as AZND2, the estimated Jacobian matrix is obtained. It is noteworthy that in the simulation phase of this chapter, $\dot{\mathbf{p}}(t)$ and $\ddot{\mathbf{p}}(t)$ are readily available. However, in practical experiments, acquiring these values remains a significant challenge. Therefore, when implementing this method in real-world applications, filters or differentiators may be employed to estimate the actual velocity and acceleration of the end-effector [90].

Subsequently, instead of numerically computing the pseudo-inverse of the Jacobian matrix, as done in (5.29), the ZND design formula is applied for the third time [38]. Given that the right pseudo-inverse $J^{\dagger}$ satisfies $J^{\dagger}(t)J(t)J^{\mathrm{T}}(t) = J^{\mathrm{T}}(t)$, a matrix-valued error function is defined as follows:

$$\Pi(t) = W(t)J(t)J^{\mathrm{T}}(t) - J^{\mathrm{T}}(t), \tag{5.32}$$

where $W(t)$ denotes the pseudo-inverse of $J(t)$, which is to be determined. Applying the ZND design formula once more yields

$$\dot{\Pi}(t) = -\gamma \mathcal{F}(\Pi(t)), \tag{5.33}$$

where $\mathcal{F}$ represents a matrix-to-matrix array of activation functions. The power sigmoid activation function is employed in this chapter. Substituting (5.32) into (5.33) leads to

$$\begin{aligned}&\dot{W}(t)J(t)J^{\mathrm{T}}(t) + W(t)\left(\dot{J}(t)J^{\mathrm{T}}(t) + J(t)\dot{J}^{\mathrm{T}}(t)\right) - \dot{J}^{\mathrm{T}}(t)\\ &= -\gamma \mathcal{F}\left(W(t)J(t)J^{\mathrm{T}}(t) - J^{\mathrm{T}}(t)\right).\end{aligned} \tag{5.34}$$

The above equation can be rewritten in the following explicit form:

$$\begin{aligned}\dot{W}(t) = \Big(&-\gamma \mathcal{F}\left(W(t)J(t)J^{\mathrm{T}}(t) - J^{\mathrm{T}}(t)\right) - W(t)\left(\dot{J}(t)J^{\mathrm{T}}(t) + J(t)\dot{J}^{\mathrm{T}}(t)\right)\\ &+\dot{J}^{\mathrm{T}}(t)\Big)\left(J(t)J^{\mathrm{T}}(t)\right)^{-1},\end{aligned} \tag{5.35}$$

which is termed AZND3.

As observed in (5.35), the equation still contains an inverse operation. To address this, an additional definition is proposed:

$$V(t) = \left(J(t)J^{\mathrm{T}}(t)\right)^{-1}. \tag{5.36}$$

To compute $V(t)$, the following matrix-valued error function is defined:

$$\Xi(t) = J(t)J^{\mathrm{T}}(t) - V^{-1}(t). \tag{5.37}$$

Utilizing the ZND design formula for the fourth time and substituting (5.37) into it yield

$$\dot{\Xi}(t) = -\gamma \Xi(t). \tag{5.38}$$

According to Liao and Zhang [91], for a time-varying non-singular matrix, we have

$$\dot{V}^{-1}(t) = -V^{-1}(t)\dot{V}(t)V^{-1}(t). \tag{5.39}$$

Substituting (5.37) into (5.38) and considering (5.39), the following AZND4 model is obtained:

$$\dot{V}(t) = -\gamma\left(V(t)J(t)J^{\mathrm{T}}(t)V(t) - V(t)\right) - V(t)\left(\dot{J}(t)J^{\mathrm{T}}(t) + J(t)\dot{J}^{\mathrm{T}}(t)\right)V(t). \tag{5.40}$$

By solving the above dynamic equation, $V(t)$ is determined. Consequently, the dynamic equation associated with AZND3 is reformulated as:

$$\begin{aligned}\dot{W}(t) = \Big(& -\gamma\mathcal{F}\left(W(t)J(t)J^{\mathrm{T}}(t) - J^{\mathrm{T}}(t)\right) \\ & -W(t)\left(\dot{J}(t)J^{\mathrm{T}}(t) + J(t)\dot{J}^{\mathrm{T}}(t)\right) + \dot{J}^{\mathrm{T}}(t)\Big)V(t),\end{aligned} \tag{5.41}$$

where $W(t)$ represents the pseudo-inverse $J^{\dagger}(t)$ of the Jacobian matrix $J(t)$.

Additionally, the dynamic equation associated with AZND1 is rewritten as

$$\ddot{\mathbf{u}}(t) = W(t)\Big(\ddot{\mathbf{p}}_{\mathrm{d}}(t) - \dot{J}(\mathbf{u}(t))\dot{\mathbf{u}}(t) + \gamma\big(\,(\dot{\mathbf{p}}_{\mathrm{d}}(t) - \dot{\mathbf{p}}(t)) + (\mathbf{p}_{\mathrm{d}}(t) - \mathbf{p}(t))\,\big)\Big). \tag{5.42}$$

In summary, the ACQN method is established by applying the ZND design formula four times. These neural networks are essentially modeled as systems of ordinary differential equations. Thus, implementation involves solving the system of ordinary differential Eqs. (5.31), (5.40), (5.41), and (5.42).

5.4.2 ADQN Method

To develop the ADQN model, a transition from a continuous framework to a discrete one is made. This transformation enhances compatibility with digital platforms and facilitates implementation. The ACQN method involves four continuous-time ordinary differential equations. To discretize these equations, four different formulas are employed: (1) the TZD formula [92–94], (2) the Verlet method [87], (3) the Euler forward formula, and (4) the Euler backward formula.

Let r_i denote the value of a time-varying variable $r(t)$ at the discrete-time instant $t_i = i\tau$, with τ representing the sampling interval.

The 4-instant TZD formula is given by

$$\dot{r}_i = \frac{2r_{i+1} - 3r_i + 2r_{i-1} - r_{i-2}}{2\tau} + O(\tau^2), \tag{5.43}$$

where $O(\tau^2)$ is a vector whose entries are of order $O(\tau^2)$.

The 3-instant backward finite difference formula is given by

$$\dot{r}_i = \frac{3r_i - 4r_{i-1} + r_{i-2}}{2\tau} + O(\tau^2). \tag{5.44}$$

The Euler forward formula is given by

$$\dot{r}_i = \frac{r_{i+1} - r_i}{\tau} + O(\tau). \tag{5.45}$$

The Euler backward formula is given by

$$\dot{r}_i = \frac{r_i - r_{i-1}}{\tau} + O(\tau). \tag{5.46}$$

A key component of this discretization process is the employment of the Verlet integration method to precisely transform actuation accelerations into actuation magnitude. This is achieved using the following equation:

$$\mathbf{u}_{i+1} \leftarrow 2\mathbf{u}_i - \mathbf{u}_{i-1} + \tau^2\ddot{\mathbf{u}}_i, \tag{5.47}$$

where $\leftarrow$ denotes the computational assignment operator, $\mathbf{u}_i$ is the actuation magnitude at the current time step, $\mathbf{u}_{i-1}$ is the actuation magnitude at the previous time step, $\ddot{\mathbf{u}}_i$ is the acceleration at the current time step, and τ represents the sampling interval. This equation calculates the position at the subsequent time step using the current and previous positions and the actuation acceleration.

To calculate the variables at the initial two steps, the following formula is employed to estimate the position:

$$\mathbf{u}_{i+1} \leftarrow \mathbf{u}_i + \dot{\mathbf{u}}_i\tau + 0.5\ddot{\mathbf{u}}_i\tau^2, \tag{5.48}$$

where $\dot{\mathbf{u}}_i$ is the actuation velocity, estimated using (5.46), $\ddot{\mathbf{u}}_i$ is the actuation acceleration, and τ is the sampling interval.

To compute J_i, W_i, and V_i at the initial two steps, (5.45) is adopted accordingly:

$$J_{i+1} \leftarrow J_i + \tau\dot{J}_i, \tag{5.49}$$

$$W_{i+1} \leftarrow W_i + \tau\dot{W}_i, \tag{5.50}$$

$$V_{i+1} \leftarrow V_i + \tau\dot{V}_i. \tag{5.51}$$

The derivatives $\dot{J}_i$, $\dot{W}_i$, and $\dot{V}_i$ are calculated by sampling (5.31), (5.40), and (5.41) at time t_i, respectively, as follows:

$$\dot{J}_i = \left(\ddot{\mathbf{p}}_i - J_i\ddot{\mathbf{u}}_i + \frac{h}{\tau}\left(\dot{\mathbf{p}}_i - J_i\dot{\mathbf{u}}_i\right)\right)\frac{\dot{\mathbf{u}}_i^{\mathrm{T}}}{\dot{\mathbf{u}}_i^{\mathrm{T}}\dot{\mathbf{u}}_i}, \tag{5.52}$$

$$\dot{W}_i = \left(-\frac{h}{\tau}\mathcal{F}\left(W_iJ_iJ_i^{\mathrm{T}} - J_i^{\mathrm{T}}\right) - W_i\left(\dot{J}_iJ_i^{\mathrm{T}} + J_i\dot{J}_i^{\mathrm{T}}\right) + \dot{J}_i^{\mathrm{T}}\right)V_i, \tag{5.53}$$

$$\dot{V}_i = -\frac{h}{\tau}\left(V_iJ_iJ_i^{\mathrm{T}}V_i - V_i\right) - V_i\left(\dot{J}_iJ_i^{\mathrm{T}} + J_i\dot{J}_i^{\mathrm{T}}\right)V_i, \tag{5.54}$$

where $h = \lambda\tau$. The step size h is vital in determining the stability of discretized acceleration-level neurodynamics. According to the TZD framework, the effective range of h for ensuring numerical stability is (0, 1) [95].

By adopting the Euler backward formula (5.46), $\dot{\mathbf{u}}_i$ is calculated as

$$\dot{\mathbf{u}}_i \leftarrow \frac{\mathbf{u}_i - \mathbf{u}_{i-1}}{\tau}. \tag{5.55}$$

$\ddot{\mathbf{u}}_i$ is computed by sampling (5.42):

$$\ddot{\mathbf{u}}_i = W_i \left(\ddot{\mathbf{p}}_{\mathrm{d},i} - \dot{J}_i \dot{\mathbf{u}}_i + \delta \left((\dot{\mathbf{p}}_{\mathrm{d},i} - \dot{\mathbf{p}}_i) + (\mathbf{p}_{\mathrm{d},i} - \mathbf{p}_i) \right) \right). \tag{5.56}$$

Remark 5.1: To avoid the singularity problem, a damping factor λ, which enhances numerical stability in the pseudo-inverse computation, has been introduced. The updated equation introduces a regularization term in the denominator to prevent singularities:

$$\dot{J}_i = \left(\ddot{\mathbf{p}}_i - J_i \ddot{\mathbf{u}}_i + \frac{h}{\tau} (\dot{\mathbf{p}}_i - J_i \dot{\mathbf{u}}_i) \right) \frac{\dot{\mathbf{u}}_i^{\mathrm{T}}}{\dot{\mathbf{u}}_i^{\mathrm{T}} \dot{\mathbf{u}}_i + \lambda^2}. \tag{5.57}$$

By incorporating λ^2, this adjustment guarantees that the denominator remains strictly positive, eliminating the risk of division by zero.

To enhance accuracy in subsequent steps of the task, the discrete-time states of the variables are determined using the 4-instant TZD formula (5.43) and the Verlet method (5.47):

$$\mathbf{u}_{i+1} \leftarrow \mathbf{u}_i + \dot{\mathbf{u}}_i \tau + \frac{1}{2} \ddot{\mathbf{u}}_i \tau^2, \tag{5.58}$$

$$J_{i+1} \leftarrow \frac{3}{2} J_i - J_{i-1} + \frac{1}{2} J_{i-2} + \tau \dot{J}_i, \tag{5.59}$$

$$W_{i+1} \leftarrow \frac{3}{2} W_i - W_{i-1} + \frac{1}{2} W_{i-2} + \tau \dot{W}_i, \tag{5.60}$$

$$V_{i+1} \leftarrow \frac{3}{2} V_i - V_{i-1} + \frac{1}{2} V_{i-2} + \tau \dot{V}_i. \tag{5.61}$$

It is evident that the ADQN method is entirely inverse free. The procedure of the ADQN method is presented in Algorithm 2.

Finally, the ILR3S [88] formula is employed to discretize the joint acceleration into joint position. By applying the ILR3S formula, the following algorithm is obtained:

$$\mathbf{u}_{i+1} \leftarrow \frac{3}{4} \mathbf{u}_i + \frac{1}{8} \mathbf{u}_{i-1} + \frac{1}{8} \mathbf{u}_{i-2} + \tau \left(\frac{71}{192} \dot{\mathbf{u}}_{i+1} + \frac{173}{192} \dot{\mathbf{u}}_i - \frac{5}{192} \dot{\mathbf{u}}_{i-1} + \frac{5}{64} \dot{\mathbf{u}}_{i-2} \right), \tag{5.62}$$

$$\dot{\mathbf{u}}_{i+1} \leftarrow 2\dot{\mathbf{u}}_i - \frac{3}{2} \dot{\mathbf{u}}_{i-1} + \frac{1}{2} \dot{\mathbf{u}}_{i-2} + \tau \left(\frac{35}{24} \ddot{\mathbf{u}}_i - \frac{5}{3} \ddot{\mathbf{u}}_{i-1} + \frac{17}{24} \ddot{\mathbf{u}}_{i-2} \right). \tag{5.63}$$

Through a slight modification of the ILR3S formula, $\dot{\mathbf{u}}_i$ is calculated using (5.55) rather than the ZND formula, while $\ddot{\mathbf{u}}_i$ is calculated using (5.56). Thus, the actuation acceleration is discretized into the actuation magnitude using a modified ILR3S formula.

Algorithm 2: Procedure of ADQN method.

Initialize: $\mathbf{u}_0, J_0, W_0, V_0, \dot{\mathbf{u}}_0, \ddot{\mathbf{u}}_0$
Set: T, h, τ, $\mathbf{p}_\mathrm{d}$
for $i = 0 : \frac{T}{\tau} - 1$ **do**
  Compute:

  1. First derivative of Jacobian matrix, $\dot{J}_i$, via (5.52)
  2. Jacobian pseudo-inverse, $\dot{W}_i$ and $\dot{V}_i$, via (5.53) and (5.54), respectively
  3. Actuation velocity, $\dot{\mathbf{u}}_i$, via (5.55)
  4. Actuation acceleration, $\ddot{\mathbf{u}}_i$, via (5.56)

  if $i \leq 2$ **then**
    Compute future state:

  1. Actuation position, $\mathbf{u}_{i+1}$, via (5.48)
  2. Jacobian matrix, J_{i+1}, via (5.49)
  3. Jacobian pseudo-inverse, W_{i+1} and V_{i+1}, via (5.50) and (5.51), respectively

  else
    Compute future state: $\mathbf{u}_{i+1}, J_{i+1}, W_{i+1}$, and V_{i+1} via (5.58), (5.59), (5.60), and (5.61), respectively.
  end if
end for

5.5 Verification of DQND4

In this section, simulations and physical experiments based on continuum robots are carried out to validate the DQND4 method.

5.5.1 Simulations

To validate the DQND4 model, simulations are conducted on a two-segment continuum robot, each segment equipped with three actuators. The tracking task duration is set to $T = 20$ s. The initial estimation for the Jacobian matrix is established by incrementally adjusting each actuator independently by Δc to measure the displacement of the end-effector, $\Delta\mathbf{k}_i$. This approach enables the initialization of the Jacobian matrix columns as $J_i(0) = \Delta\mathbf{k}_i/\Delta c$, with X initialized as the pseudo-inverse of the Jacobian matrix, i.e., $X(0) = J^\dagger(0)$. For this simulation, the sampling interval τ is chosen as $\tau = 0.001$ s, and the parameter h is set to 0.1.

The performance of the DQND4 model for trajectory tracking of a pentagram path is illustrated in Figure 5.2. As illustrated in Figure 5.2a, the end-effector of the continuum robot closely follows the pentagram path, with minimal deviation between the desired path and actual trajectory. Position errors $\mathbf{e} = [e_x, e_y, e_z]^\mathrm{T} = \mathbf{p}_\mathrm{d} - \mathbf{p}$ along the x-, y-, and z-axis are shown in Figure 5.2b, where the root-mean-square error (RMSE) for the pentagram path is less than 1×10^{-10} m.

5.5.2 Experiments

To experimentally validate the DQND4 control algorithm, a physical tendon-driven continuum robot is utilized. In this experiment, the DQND4 algorithm is applied to guide the continuum robot along a heart-shaped trajectory, with snapshots shown in Figure 5.3. The total duration for the path tracking task is set to 80 s, with a sampling interval τ of 0.1 s, and the parameter h configured to 0.05. As depicted in Figure 5.4, the RMSE achieved is 9.2×10^{-4} m, indicating that the robot effectively completes the tracking task with high accuracy.

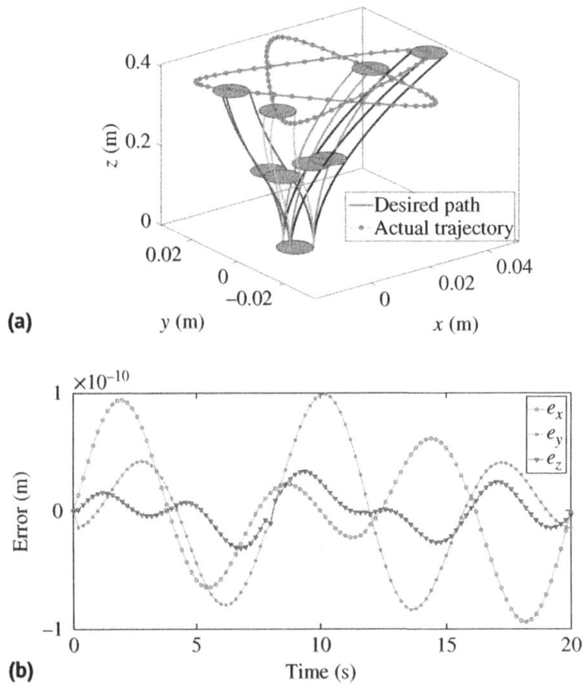

Figure 5.2 Pentagram path tracking using DQND4 model with $\tau = 0.001$ s and $h = 0.1$: (a) configurations of robot; (b) tracking errors.

Figure 5.3 Heart path tracking experiment on continuum robot platform.

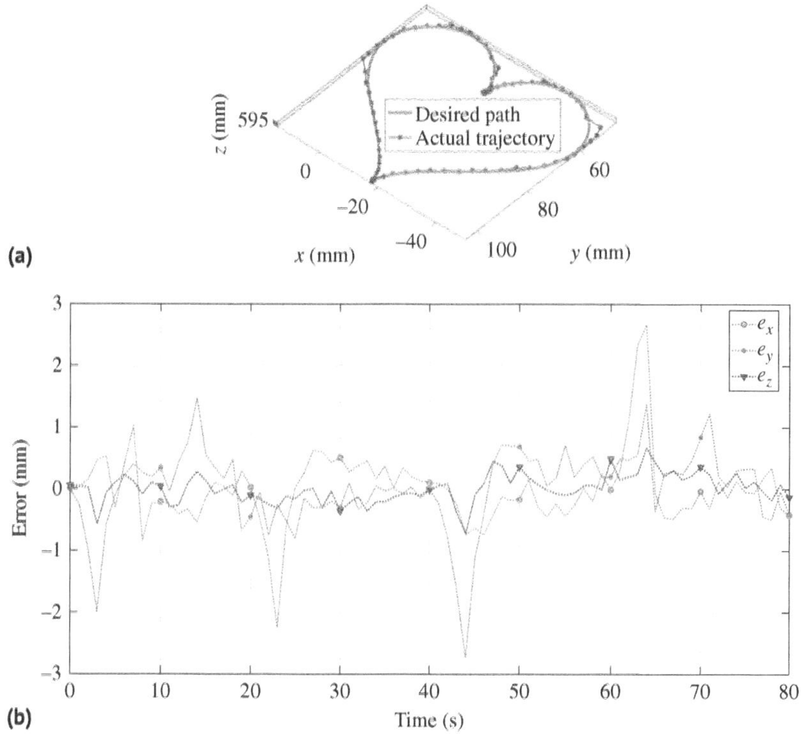

Figure 5.4 Experiment results of heart path tracking task aided by DQND4 method: (a) trajectory comparison; (b) tracking errors.

5.6 Verification of ADQN

This section provides simulation results to validate the effectiveness of the ADQN control method for trajectory tracking in continuum robots.

5.6.1 Setting

In simulations, a continuum robot manipulator consisting of two segments, each actuated by three independent actuators, is simulated. This setup balances flexibility and control complexity. The trajectory tracking task lasts for $T = 20$ s, allowing a comprehensive assessment of the robot's performance. The Jacobian matrix $J(0)$ is initialized by measuring the displacement $\Delta\mathbf{p}_i$ of the end-effector by independently perturbing each actuator with a small increment Δc. The ith column of the Jacobian is then computed as $J_i(0) = \Delta\mathbf{p}_i/\Delta c$. The pseudo-inverse of the Jacobian matrix, $W(0)$, is computed as the Moore–Penrose inverse of $J(0)$, i.e., $W(0) = J^\dagger(0)$. The sampling interval τ is set to 0.001 s to ensure a high sampling frequency for precise control updates. The step size h, which influences the convergence rate of the ADQN algorithm, is selected as $h = 0.1$, based on theoretical insights, to balance computational efficiency, accuracy, and system stability.

5.6.2 Results

The performance of the ADQN method is evaluated through simulations where the continuum robot tracks three distinct trajectories: pentagram path, lambda path, and rhodonea curve. These trajectories are chosen to test the method's ability to handle both smooth and complex paths, featuring challenges like sharp turns, sudden directional changes, and intricate patterns.

For the pentagram path, a star-shaped trajectory with five vertices and intersecting segments, Figure 5.5 shows the tracking results. The end-effector successfully navigates sharp corners and intersections while closely following the desired path, achieving an exceptionally low RMSE of 1.03×10^{-6} m at a sampling interval of $\tau = 0.001$ s.

The lambda path evaluation, featuring sharp turns and abrupt directional changes, exhibits similar precision. As shown in Figure 5.6, the end-effector accurately follows the challenging path with an RMSE of 4.98×10^{-6} m.

For the flower-like rhodonea curve, Figure 5.7 reveals the high-fidelity tracking capability of the ADQN method, achieving an RMSE of 5.77×10^{-7} m.

Table 5.1 confirms that tracking accuracy improves significantly with smaller sampling intervals, highlighting the importance of high-frequency sampling for complex trajectories.

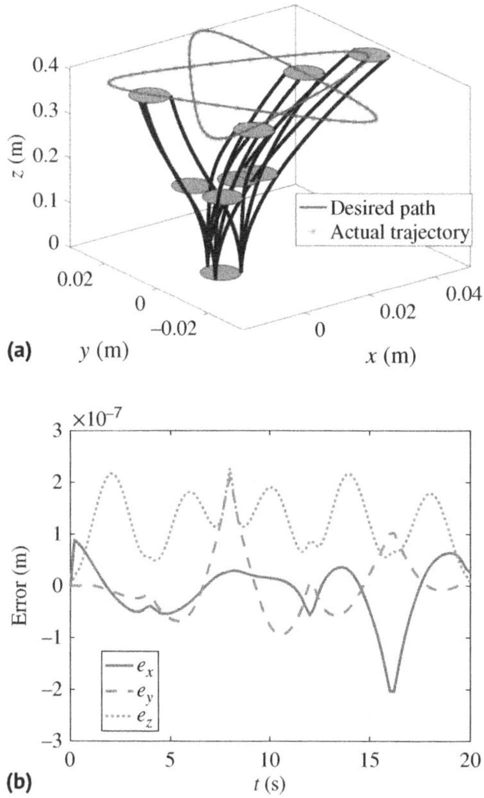

Figure 5.5 Pentagram path tracking using ADQN with $\tau = 0.001$ s and $h = 0.1$: (a) tracking process; (b) tracking errors.

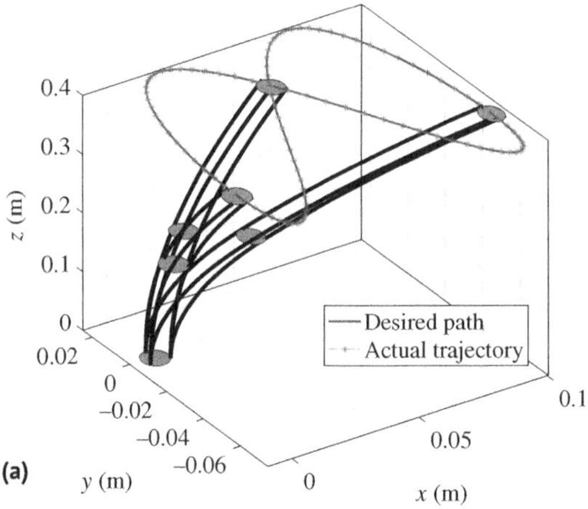

(a)

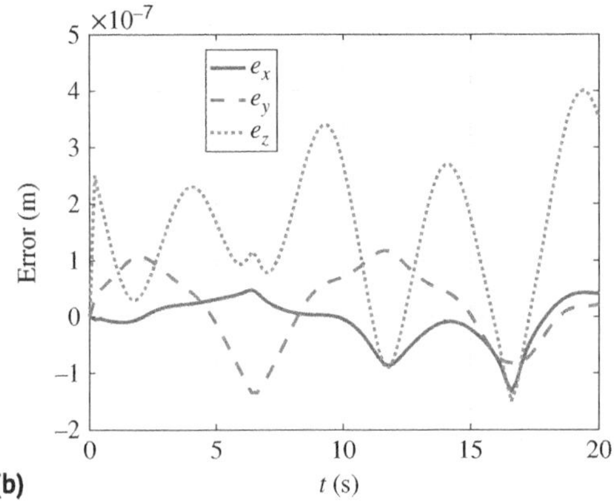

(b)

Figure 5.6 Lambda path tracking using ADQN with $\tau = 0.001$ s and $h = 0.1$: (a) tracking process; (b) tracking errors.

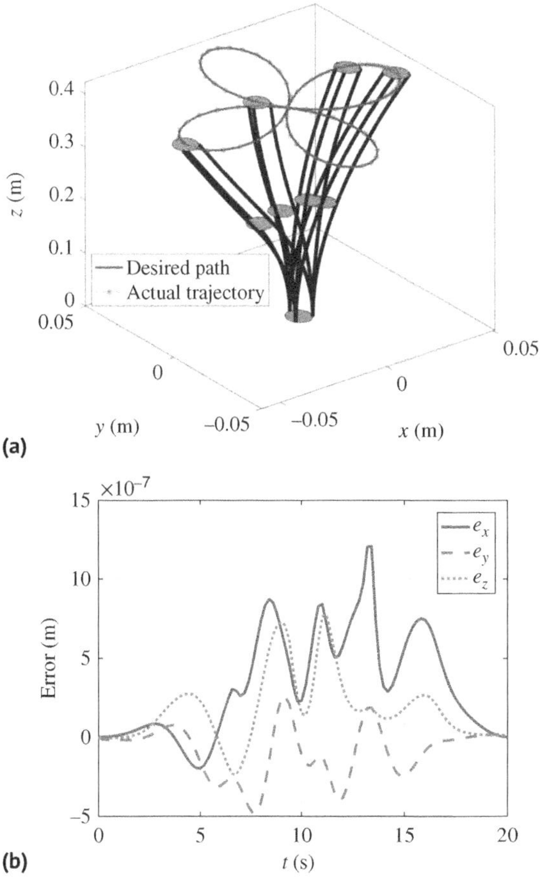

Figure 5.7 Rhodonea path tracking using ADQN with $\tau = 0.001$ s and $h = 0.1$: (a) tracking process; (b) tracking errors.

Table 5.1 RMSE comparison for various trajectories at different sampling intervals τ.

τ (s)	Rhodonea (m)	Pentagram (m)	Lambda (m)
0.1	7.02×10^{-3}	1.97×10^{-3}	5.20×10^{-3}
0.01	5.85×10^{-5}	3.62×10^{-5}	1.58×10^{-4}
0.001	5.77×10^{-7}	1.03×10^{-6}	4.98×10^{-6}

5.7 Chapter Summary

This chapter presents control strategies for continuum robots through the development of CQND and DQND4 models. The CQND-based control system was designed for tracking control of continuum robots, utilizing the solver for inverse kinematics, the estimator for Jacobian matrix, and the matrix pseudo-inverse constructed based on the ZND design formula. To enhance practical applicability, this system was discretized, resulting in the DQND4 model, which employs four distinct discretization formulas. Besides, this chapter introduces an acceleration-level control approach for continuum robots through the ACQN framework, representing a significant advancement from conventional velocity-level control strategies. The proposed method improves precision, adaptability, and operational reliability by enabling direct acceleration modulation. The discretization of ACQN using both the Verlet method and modified ILR3S formula yields the ADQN model, which maintains an inverse-free formulation while enhancing trajectory tracking and overall system performance. Simulations and experiments confirmed the effectiveness and accuracy of the methods.

Part V

Variants of Zeroing Neurodynamics Methods

6

Nonlinearly Activated Zeroing Neurodynamic Control Methods

6.1 Introduction

Model-free control algorithms for continuum robots have been extensively developed [96]. For example, Thuruthel et al. [19] proposed a closed-loop kinematic controller based on a machine learning approach. Li et al. [51] introduced a model-free control method utilizing an adaptive Kalman filter for continuum robots to address the trajectory tracking issue. Despite advancements in model-free control for continuum robots, prior studies are limited by their low accuracy and inadequate robustness.

In recent years, significant progress has been made in the research of neural networks and artificial intelligence [97–99]. Compared with the feed-forward neural networks employed in the above methods, a novel recurrent neural network method, termed as zeroing neurodynamics (ZND) [27], represents a superior option for controlling continuum robots since it is designed for time-varying problems and does not require the pre-training process. For example, Chen and Zhang [100] developed a Jacobian-matrix-adaption method to solve the tracking control problem of the PUMA 560 robot manipulator based on ZND and achieved excellent results. Xie et al. [101] proposed an acceleration-level data-driven repetitive motion planning scheme for the kinematic control of robots with unknown structure. The aforementioned schemes are implemented under ideal conditions or consider only random noise. However, various types of noise and external disturbances are always existed in the real world. Therefore, this motivates the development of a more robust, noise-tolerant, and model-free approach based on ZND to solve the kinematic control problem of continuum robots. In addition, achieving finite-time convergence is also essential for the effective application of the control scheme [102, 103].

In light of the advantages of ZND, various modified versions of ZND have been proposed and investigated to improve convergence and robustness. For the original ZND, a linear activation function (LAF) is employed. To enhance the performance of the original ZND, various nonlinear activation functions have been proposed and examined, as different activation functions influence stability and convergence differently. For instance, the combined activation function (CAF-ZND) [42] and integration-enhanced linear activation function (IE-ZND) [104] are utilized to improve the robustness of ZND under noise disturbances. A versatile activation function (VAF-ZND) is introduced to achieve finite-time convergence and to decrease the impact of noise [45]. However, the solution error of CAF-ZND and VAF-ZND cannot converge to zero when subjected to constant noise disturbances, whereas IE-ZND performs poorly in the presence of random noise. To address these limitations, an integration-enhanced noise-tolerant activation function is introduced for the ZND model (IENT-ZND) [105].

This chapter introduces several noise-tolerant and convergence-enhanced ZND models and a dual ZND scheme for the kinematic control of continuum robots with unknown models. Several noises, including constant noise, time-varying noise, and random noise, are considered in simulations and physical experiments to validate and compare the effectiveness and capability of different ZND models and the dual ZND scheme.

6.2 Variants of ZND Models

For a dynamic system, let $z(t)$ represent the actual state and $z^*(t)$ denote the desired state. According to the ZND design method [27], an indefinite error function is initially defined as follows:

$$e(t) = z^*(t) - z(t). \tag{6.1}$$

Subsequently, the following formula is selected as the time derivative of the error function to ensure that the error converges to zero:

$$\dot{e}(t) = -\beta\phi(e(t)), \tag{6.2}$$

where $\beta > 0$ is the design parameter of the ZND model (6.2) for adjusting the convergence rate of the model, and ϕ is the activation function. Finally, substituting the error function into (6.2) and solving the differential equation will yield the solution to the problem.

The comparisons and formulations of activation functions for different ZND models are summarized in Table 6.1, where $\text{sgn}(\cdot)$ represents the sign function. Specially, the convergence and robustness analyses of these ZND models can be found in the literature. As an example, the convergence and robustness analysis regarding the IENT-ZND model under different conditions is elaborated below.

Table 6.1 Comparisons and formulations of different ZND models.

Models	Activation function
LAF-ZND	$\phi(x) = x$
CAF-ZND	$\phi(x) = k_1(\lvert x\rvert^{\rho} + \lvert x\rvert^{\frac{1}{\rho}})\text{sgn}(x) + k_2 x$ with $k_1 > 0$, $k_2 > 0$ and $\rho \in (0, 1)$
IE-ZND	$\phi(x) = x + \int_0^t x(\tau)\text{d}\tau$
VAF-ZND	$\phi(x) = (\lvert x\rvert^{\eta} + \lvert x\rvert^{w})\text{sgn}(x) + x + \text{sgn}(x)$ with $\eta \in (0, 1)$ and $w > 1$
IENT-ZND	$\phi(x) = (\lvert x\rvert^{\mu} + \lvert x\rvert^{\frac{1}{\mu}} + \lvert x\rvert + \psi)\text{sgn}(x) + \int_0^t x(\tau)\text{d}\tau$ with $\mu \in (0, 1)$ and $\psi > 0$

6.2.1 Analysis Under Noise-free Condition

In this section, we analyze the finite-time convergence of the IENT-ZND model in the absence of noise by the following theorem.

Theorem 6.1 *When the system dynamics are described by the IENT-ZND model, the system state $z(t)$ converges to the desired state $z^*(t)$ in finite-time $t_{\max}$, where*

$$t_{\max} \leq \frac{1}{\beta(1-\mu)} \ln\left(1 + \lvert e(0)\rvert^{1-\mu}\right). \tag{6.3}$$

Proof: Firstly, based on the design formula and activation function associated with the IENT-ZND model, we have

$$\dot{e}(t) = -\beta\left(\lvert e(t)\rvert^{\mu} + \lvert e(t)\rvert^{\frac{1}{\mu}} + \lvert e(t)\rvert + \psi\right)\text{sgn}(e(t)) - \beta\int_0^t e(\tau)\text{d}\tau. \tag{6.4}$$

Then, we have the following three cases.

- $e(0) > 0$. Suppose that $e(t)$ converges to 0 at the time instant $t_{\max}$. In this case, when $t \in [0, t_{\max})$, we rewrite (6.4) as

$$\dot{e}(t) = -\beta\left(e^{\mu}(t) + e^{\frac{1}{\mu}}(t) + e(t) + \psi\right) - \beta\int_0^t e(\tau)\text{d}\tau. \tag{6.5}$$

Since $\beta > 0$ and $e(t) > 0$, when $t \in [0, t_{\max})$, one further gets

$$\begin{aligned}\dot{e}(t) &\leq -\beta\left(e^{\mu}(t) + e^{\frac{1}{\mu}}(t) + e(t) + \psi\right) \\ &< -\beta\left(e^{\mu}(t) + e(t)\right).\end{aligned} \tag{6.6}$$

That is

$$\frac{\mathrm{d}e(t)}{\mathrm{d}t} < -\beta\left(e^{\mu}(t) + e(t)\right), \tag{6.7}$$

$$\frac{\mathrm{d}e^{1-\mu}(t)}{\mathrm{d}t} < -\beta\left(1 + e^{1-\mu}(t)\right)(1-\mu).$$

Then, the following inequality is attained:

$$\mathrm{d}t < -\frac{1}{\beta(1-\mu)\left(1 + e^{1-\mu}(t)\right)}\mathrm{d}e^{1-\mu}(t). \tag{6.8}$$

Integrating (6.8) from $t = 0$ to $t = t_{\max}$ leads to

$$\int_0^{t_{\max}} \mathrm{d}t < \int_{e^{1-\mu}(0)}^{0} -\frac{\mathrm{d}e^{1-\mu}(t)}{\beta(1-\mu)\left(1 + e^{1-\mu}(t)\right)}, \tag{6.9}$$

$$t\Big|_0^{t_{\max}} < -\frac{1}{\beta(1-\mu)}\ln\left(1 + e^{1-\mu}(t)\right)\Big|_{e^{1-\mu}(0)}^{0},$$

$$t_{\max} < \frac{1}{\beta(1-\mu)}\ln\left(1 + e^{1-\mu}(0)\right).$$

Therefore, inequality (6.3) holds true.

- $e(0) < 0$. Suppose that $e(t)$ converges to 0 at the time instant $t_{\max}$. Since $e(t) < 0$, when $t \in [0, t_{\max})$, we rewrite (6.4) as

$$\dot{e}(t) = \beta\left(|e(t)|^{\mu} + |e(t)|^{\frac{1}{\mu}} + |e(t)| + \psi\right) + \beta\int_0^t |e(\tau)|\mathrm{d}\tau. \tag{6.10}$$

Then, one further gets

$$\dot{e}(t) \geq \beta\left(|e(t)|^{\mu} + |e(t)|^{\frac{1}{\mu}} + |e(t)| + \psi\right) \tag{6.11}$$

$$> \beta\left(|e(t)|^{\mu} + |e(t)|\right).$$

That is

$$\frac{\mathrm{d}|e(t)|}{\mathrm{d}t} < -\beta\left(|e(t)|^{\mu} + |e(t)|\right), \tag{6.12}$$

$$\frac{\mathrm{d}|e(t)|^{1-\mu}}{\mathrm{d}t} < -\beta\left(1 + |e(t)|^{1-\mu}\right)(1-\mu).$$

Then, the following inequality is attained:

$$\mathrm{d}t < -\frac{\mathrm{d}|e(t)|^{1-\mu}}{\beta(1-\mu)\left(1 + |e(t)|^{1-\mu}\right)}. \tag{6.13}$$

Integrating (6.13) from $t = 0$ to $t = t_{\max}$ leads to

$$\begin{aligned}\int_0^{t_{\max}} \mathrm{d}t &< \int_{|e(0)|^{1-\mu}}^{0} \frac{-\mathrm{d}|e(t)|^{1-\mu}}{\beta(1-\mu)\left(1+|e(t)|^{1-\mu}\right)},\\ t\Big|_0^{t_{\max}} &< \frac{-1}{\beta(1-\mu)} \ln\left(1+|e(t)|^{1-\mu}\right)\Big|_{|e(0)|^{1-\mu}}^{0},\\ t_{\max} &< \frac{1}{\beta(1-\mu)} \ln\left(1+|e(0)|^{1-\mu}\right).\end{aligned} \tag{6.14}$$

Therefore, inequality (6.3) holds true.

- $e(0) = 0$. In this case, we can already conclude that $t_{\max} = 0$, which means that the inequality (6.3) in the theorem holds true.

Based on the above results, one can conclude that the error between the desired state and the actual state of the dynamic system synthesized by the IENT-ZND model converges to zero in finite-time $t_{\max}$. The proof is thus completed.

6.2.2 Analysis Under Noise-polluted Condition

When disturbed by noise, the IENT-ZND model is rewritten as

$$\dot{e}(t) = -\beta\phi(e(t)) + \eta(t), \tag{6.15}$$

where $\eta(t)$ is the time-varying noise. In this section, the finite-time convergence of the IENT-ZND model under the disturbance of noise is analyzed by the following theorem.

Theorem 6.2 *When the system dynamics is described by the noise-polluted IENT-ZND model, where the bounded noise is denoted as $|\eta(t)| \leq \delta$, the system state $z(t)$ converges to the desired state $z^*(t)$ in finite-time $t_{\max}$, where*

$$t_{\max} \leq \frac{1}{\beta(1-\mu)} \ln\left(1+|e(0)|^{1-\mu}\right). \tag{6.16}$$

Proof: Firstly, based on the design formula and activation function associated with the noise-polluted IENT-ZND model, we have

$$\dot{e}(t) = -\beta\left(|e(t)|^{\mu} + |e(t)|^{\frac{1}{\mu}} + |e(t)| + \psi\right)\operatorname{sgn}(e(t)) - \beta\int_0^t e(\tau)\mathrm{d}\tau + \eta(t). \tag{6.17}$$

Then, we have the following two cases:

- $e(t) \geq 0$. Suppose that $e(t)$ converges to 0 at the time instant $t_{\max}$. In this case, when $t \in [0, t_{\max})$, we rewrite (6.17) as

$$\dot{e}(t) = -\beta\left(e^{\mu}(t) + e^{\frac{1}{\mu}}(t) + e(t) + \psi\right) - \beta \int_0^t e(\tau)\mathrm{d}\tau + \eta(t). \tag{6.18}$$

Considering $\beta\psi \geq \delta > 0$ and $|\eta(t)| < \delta$, we have

$$-\beta\psi + \eta(t) \leq 0. \tag{6.19}$$

One further gets

$$\begin{aligned}\dot{e}(t) &\leq -\beta\left(e^{\mu}(t) + e^{\frac{1}{\mu}}(t) + e(t)\right) - \beta \int_0^t e(\tau)\mathrm{d}\tau \\ &\leq -\beta\left(e^{\mu}(t) + e(t)\right).\end{aligned} \tag{6.20}$$

That is

$$\begin{aligned}\frac{\mathrm{d}|e(t)|}{\mathrm{d}t} &\leq -\beta\left(|e(t)|^{\mu} + |e(t)|\right), \\ \frac{\mathrm{d}|e(t)|^{1-\mu}}{\mathrm{d}t} &\leq -\beta\left(1 + |e(t)|^{1-\mu}\right)(1-\mu).\end{aligned} \tag{6.21}$$

Then, the following inequality is attained:

$$\mathrm{d}t \leq -\frac{\mathrm{d}|e(t)|^{1-\mu}}{\beta(1-\mu)\left(1 + |e(t)|^{1-\mu}\right)}. \tag{6.22}$$

Integrating the above inequality from $t = 0$ to $t = t_{\max}$ leads to

$$\begin{aligned}\int_0^{t_{\max}} \mathrm{d}t &\leq \int_{|e(0)|^{1-\mu}}^{0} -\frac{\mathrm{d}|e(t)|^{1-\mu}}{\beta(1-\mu)\left(1 + |e(t)|^{1-\mu}\right)}, \\ t\Big|_0^{t_{\max}} &\leq \frac{-1}{\beta(1-\mu)} \ln\left(1 + |e(t)|^{1-\mu}\right)\Big|_{|e(0)|^{1-\mu}}^{0}, \\ t_{\max} &\leq \frac{1}{\beta(1-\mu)} \ln\left(1 + |e(0)|^{1-\mu}\right).\end{aligned} \tag{6.23}$$

Therefore, the inequality (6.16) holds true.

- $e(t) < 0$. Suppose that $e(t)$ converges to 0 at time t. In this case, when $t \in [0, t_{\max})$, we rewrite (6.17) as

$$\dot{e}(t) = \beta\left(|e(t)|^{\mu} + |e(t)|^{\frac{1}{\mu}} + |e(t)| + \psi\right) + \beta \int_0^t |e(\tau)|\mathrm{d}\tau + \eta(t). \tag{6.24}$$

Considering $\beta\psi \geq \delta > 0$ and $|\eta(t)| < \delta$, we have

$$\beta\psi + \eta(t) \geq 0. \tag{6.25}$$

One further attains

$$\dot{e}(t) \geq \beta\left(|e(t)|^{\mu} + |e(t)|^{\frac{1}{\mu}} + |e(t)|\right) + \beta\int_0^t |e(\tau)|\mathrm{d}\tau \tag{6.26}$$
$$\geq \beta\left(|e(t)|^{\mu} + |e(t)|\right).$$

That is

$$\frac{\mathrm{d}|e(t)|}{\mathrm{d}t} < -\beta\left(|e(t)|^{\mu} + |e(t)|\right), \tag{6.27}$$
$$\frac{\mathrm{d}|e(t)|^{1-\mu}}{\mathrm{d}t} < -\beta\left(1 + |e(t)|^{1-\mu}\right)(1-\mu).$$

Then, the following inequality is attained:

$$\mathrm{d}t < -\frac{\mathrm{d}|e(t)|^{1-\mu}}{\beta(1-\mu)\left(1 + |e(t)|^{1-\mu}\right)}. \tag{6.28}$$

Integrating the above inequality from $t = 0$ to $t = t_{\max}$ leads to

$$\int_0^{t_{\max}} \mathrm{d}t < \int_{|e(0)|^{1-\mu}}^{0} \frac{-\mathrm{d}|e(t)|^{1-\mu}}{\beta(1-\mu)\left(1 + |e(t)|^{1-\mu}\right)}, \tag{6.29}$$
$$t\Big|_0^{t_{\max}} < \frac{-1}{\beta(1-\mu)}\ln\left(1 + |e(t)|^{1-\mu}\right)\Big|_{|e(0)|^{1-\mu}}^{0},$$
$$t_{\max} < \frac{1}{\beta(1-\mu)}\ln\left(1 + |e(0)|^{1-\mu}\right).$$

Based on the above results, when the IENT-ZND model is disturbed by bounded noise ($|\eta(t)| \leq \delta$) and the design parameters of the IENT-ZND model satisfy $\beta\psi \geq \delta$, one can conclude that the tracking error will converge to zero in finite-time $t_{\max}$. The proof is thus completed.

6.3 Method

This section details the formulation of a robust control system for continuum robots without prior knowledge of their structures or parameters.

To begin with, the kinematic equation of a continuum robot is given by

$$\mathbf{f}(\mathbf{u}(t)) = \mathbf{p}(t), \tag{6.30}$$

where $\mathbf{u}(t)$ denotes the state of actuators at time t, $\mathbf{p}(t)$ denotes the actual position of the end-effector, and $\mathbf{f}(\cdot)$ is a nonlinear mapping. By differentiating the equation with respect to t, the relationship between the change rate of actuators and the velocity of the end-effector is obtained as follows:

$$J(t)\dot{\mathbf{u}}(t) = \dot{\mathbf{p}}(t), \tag{6.31}$$

where $J(t) = \partial \mathbf{f}(\mathbf{u}(t))/\partial \mathbf{u}(t)$ is the Jacobian matrix of the continuum robot. The objective is to compute required actuation signal $\mathbf{u}(t)$ to steer the end-effector to track a reference path $\mathbf{p}_\mathrm{d}(t)$.

Then, a vector-valued indefinite error function is defined as follows to monitor the tracking process:

$$\mathbf{e}(t) = \mathbf{p}_\mathrm{d}(t) - \mathbf{p}(t) \in \mathbb{R}^n, \tag{6.32}$$

where $\mathbf{p}_\mathrm{d}(t)$ represents the desired path in the n-dimensional task space. Substituting (6.32) into the ZND model yields

$$\dot{\mathbf{p}}_\mathrm{d}(t) - \dot{\mathbf{p}}(t) = -\beta\boldsymbol{\phi}(\mathbf{p}_\mathrm{d}(t) - \mathbf{p}(t)), \tag{6.33}$$

where $\boldsymbol{\phi} \in \mathbb{R}^n$ comprises n activation functions ϕ, and $\dot{\mathbf{p}}_\mathrm{d}(t)$ denotes the desired velocity. Without loss of generality, all activation functions in $\boldsymbol{\phi}$ are set to the same. Then, the inverse kinematics controller is obtained as

$$\dot{\mathbf{u}}(t) = J^\dagger(t)\left(\dot{\mathbf{p}}_\mathrm{d}(t) + \beta\boldsymbol{\phi}(\mathbf{p}_\mathrm{d}(t) - \mathbf{p}(t))\right). \tag{6.34}$$

Since $\dot{\mathbf{p}}_\mathrm{d}(t)$ and $\mathbf{p}_\mathrm{d}(t)$ are user-defined inputs and $\mathbf{p}(t)$ can be measured by sensors, $J(t)$ needs to be determined before solving the differential Eq. (6.34). Without knowing the structural parameters of robots, $J(t)$ cannot be calculated directly based on its model. Therefore, another vector-valued indefinite error function is defined as follows:

$$\epsilon(t) = \dot{\mathbf{p}}(t) - J(t)\dot{\mathbf{u}}(t) \in \mathbb{R}^n. \tag{6.35}$$

Substituting (6.35) into the ZND model yields

$$\ddot{\mathbf{p}}(t) - \dot{J}(t)\dot{\mathbf{u}}(t) - J(t)\ddot{\mathbf{u}}(t) = -\beta\boldsymbol{\phi}(\dot{\mathbf{p}}(t) - J(t)\dot{\mathbf{u}}(t)), \tag{6.36}$$

where $\dot{\mathbf{p}}(t)$ and $\ddot{\mathbf{p}}(t)$, measured by sensors, are the derivative and the second derivative of $\mathbf{p}(t)$ respectively, $\dot{J}(t)$ is the derivative of the Jacobian matrix, $\ddot{\mathbf{u}}(t)$ is the second derivative of $\mathbf{u}(t)$. Finally, the Jacobian matrix is estimated by the following dynamic equation:

$$\dot{J}(t) = \left(\ddot{\mathbf{p}}(t) - J(t)\ddot{\mathbf{u}}(t) + \beta\boldsymbol{\phi}\left(\dot{\mathbf{p}}(t) - J(t)\dot{\mathbf{u}}(t)\right)\right)\dot{\mathbf{u}}^\dagger(t). \tag{6.37}$$

The inputs of the above model are the actual velocity $\dot{\mathbf{p}}(t)$ and acceleration $\ddot{\mathbf{p}}(t)$ of the end-effector obtained from sensory measurements.

Finally, as shown in Figure 6.1, the dual ZND control system for continuum robots composed of ZND1 and ZND2 is obtained. The inputs to the system are the desired path $\mathbf{p}_\mathrm{d}(t)$ and desired velocity $\dot{\mathbf{p}}_\mathrm{d}(t)$. ZND1 is employed to generate the signal $\mathbf{u}(t)$, which actuates the robot based on the sensory output $\mathbf{p}(t)$ and the Jacobian matrix $J(t)$ estimated by ZND2. The output of the system is the actual trajectory $\mathbf{p}(t)$ of the end-effector.

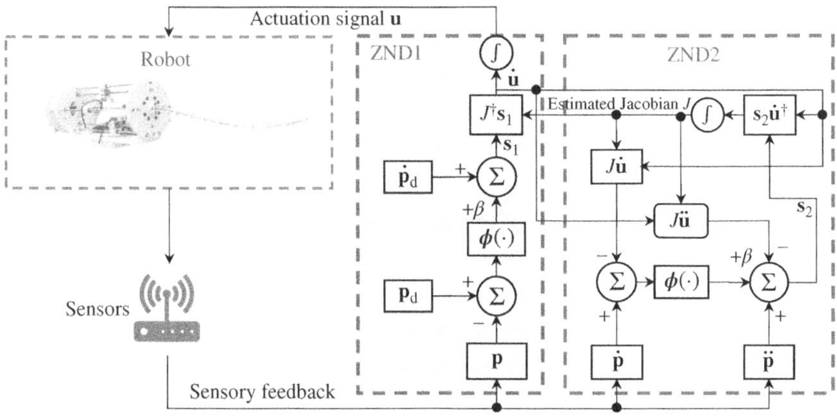

Figure 6.1 Architecture of control system for continuum robots based on dual zeroing neurodynamics (ZND).

6.4 Verification

In this section, simulations and experiments based on continuum robots are conducted to confirm the feasibility, convergence and robustness of nonlinear ZND models and the dual ZND control system in addressing the tracking control problem.

6.4.1 Simulations

To conduct simulations, a two-segment continuum robot is constructed based on the modeling method proposed by Della Santina et al. [35]. As shown in Figure 6.2, l_{ij} represents the length of the jth actuator in the ith segment and $l_{i,0}$ denotes the initial backbone length of the ith segment. Then, $\mathbf{c}_i = [\Delta_{xi}, \Delta_{yi}, \Delta_{li}]^{\mathrm{T}}$ defines the configuration parameters of the ith segment, where $\Delta_{xi} = l_{i2} - l_{i1}$, $\Delta_{yi} = l_{i4} - l_{i3}$ and Δ_{li} is the change of backbone length. Finally, the actuation state of the continuum robot is represented by $\mathbf{u}(t) = [\mathbf{c}_1^{\mathrm{T}}, \mathbf{c}_2^{\mathrm{T}}]^{\mathrm{T}}$. The task of the continuum robot is to track a double arc cycloid path within the three-dimensional task space.

6.4.1.1 Parameter Setting

The parameters involved in simulations are outlined in this part. The duration of the tracking task is defined as $T_{\mathrm{d}} = 20$ s. The initial actuation state of the continuum robot is set as $\mathbf{u}(0) = [5, 0, 0, 5, 0, 0]^{\mathrm{T}}$ mm and the initial backbone lengths are given by $l_{1,0} = l_{2,0} = 200$ mm. The design parameter of the ZND models is $\beta = 5$.

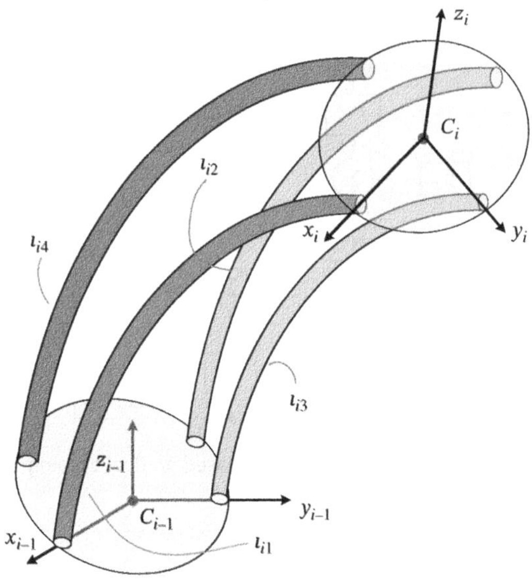

Figure 6.2 Structure of segment i of continuum robot.

Table 6.2 Parameter setting for different ZND models.

Models	Parameter values
CAF-ZND	$k_1 = 1, k_2 = 1, \rho = 0.5$
VAF-ZND	$\eta = 0.5, w = 2$
IENT-ZND	$\mu = 0.5, \psi = 1$

The parameters of different ZND models are summarized in Table 6.2. Additionally, the initial value of the Jacobian matrix $J(0)$ requires determination. By independently changing the ith element of the actuation state by an incremental amount Δu_i, the displacement $\Delta \mathbf{p}_i$ of the end-effector can be measured. Subsequently, the ith column of $J(0)$ is initialized as $J_i(0) = \Delta \mathbf{p}_i / \Delta u_i$ and the Jacobian matrix is initialized as follows:

$$J(0) = \left[\frac{\Delta \mathbf{p}_1}{\Delta u_1}, \frac{\Delta \mathbf{p}_2}{\Delta u_2}, \frac{\Delta \mathbf{p}_3}{\Delta u_3}, \frac{\Delta \mathbf{p}_4}{\Delta u_4}, \frac{\Delta \mathbf{p}_5}{\Delta u_5}, \frac{\Delta \mathbf{p}_6}{\Delta u_6} \right]. \tag{6.38}$$

The reference path in the three-dimensional task space is defined as follows:

$$\mathbf{p}_{\rm d}(t) = \begin{bmatrix} 4\varkappa\cos\left(2\pi\sin^2\left(\frac{\pi t}{2T_{\rm d}}\right)\right) + \varkappa\cos\left(6\pi\sin^2\left(\frac{\pi t}{2T_{\rm d}}\right)\right) - 5\varkappa \\ 4\varkappa\sin\left(2\pi\sin^2\left(\frac{\pi t}{2T_{\rm d}}\right)\right) + \varkappa\sin\left(6\pi\sin^2\left(\frac{\pi t}{2T_{\rm d}}\right)\right) \\ \varkappa\sin\left(2\pi\sin^2\left(\frac{\pi t}{2T_{\rm d}}\right)\right) + \frac{\varkappa}{4}\sin\left(6\pi\sin^2\left(\frac{\pi t}{2T_{\rm d}}\right)\right) \end{bmatrix} + \mathbf{p}(0), \tag{6.39}$$

where $\mathbf{p}(0) = [p_{X,0}, p_{Y,0}, p_{Z,0}]^{\rm T}$ denotes the absolute position of the desired path in the task space, specifically the initial position of the end-effector, and $\varkappa = 20$ mm is a constant to scale the desired path.

6.4.1.2 Simulation Results

This part elaborates on the simulation results of trajectory tracking for a two-segment continuum robot.

Firstly, noise is not considered to verify the feasibility to control the continuum robot with the control system using the IENT-ZND model. Figure 6.3 illustrates

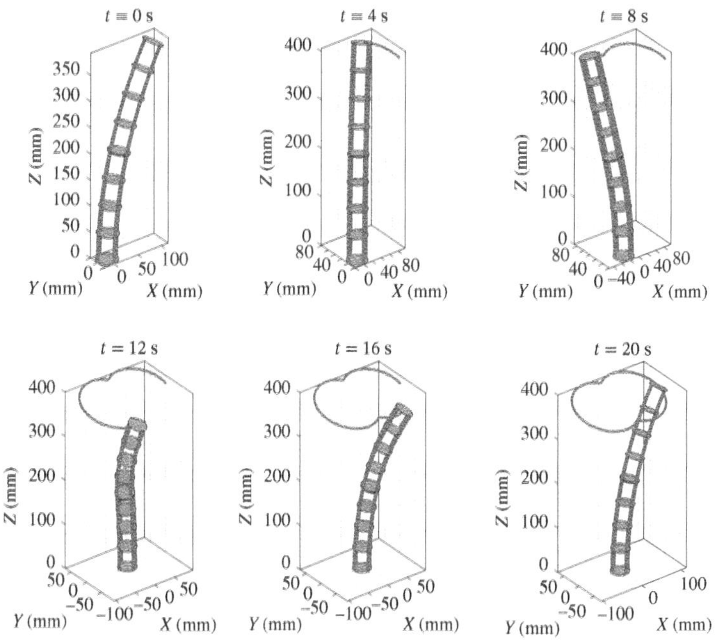

Figure 6.3 Trajectory tracking process of continuum robot for tracking double arc cycloid path.

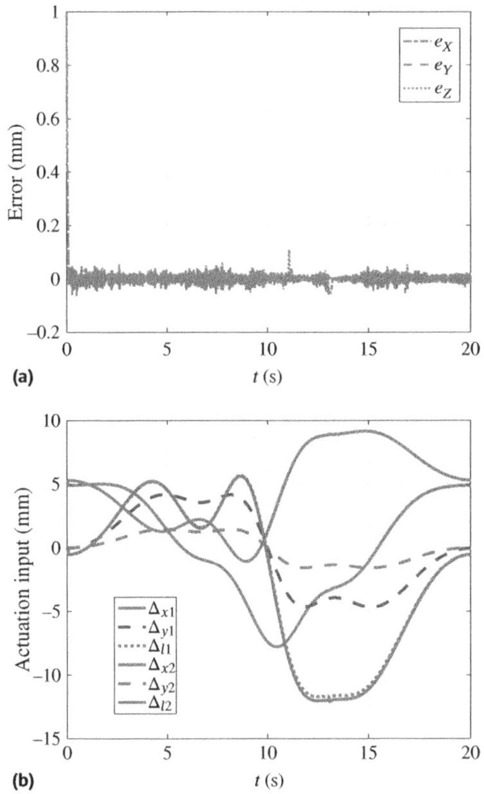

Figure 6.4 Trajectory tracking results of double arc cycloid path when using IENT-ZND-based control system without noise: (a) position errors; (b) actuation inputs.

some configurations of the continuum robot at different time instants throughout the tracking process. To validate system convergence, initial errors exist between the end-effector and the desired path. As illustrated in Figure 6.4a, the errors corresponding to different axes decrease rapidly and the root-mean-square error (RMSE) is 0.1224 mm. From Figure 6.4a, it is observed that the tracking errors almost converge to zero at $t = 0.1$ s, thereby verifying the finite-time convergence of the control system. In addition, the changes of configuration parameters during the task execution are illustrated in Figure 6.4b.

Despite the satisfactory performance of the IENT-ZND-based control scheme in the noise-free case, the robustness of the control system based on the IENT-ZND compared with other ZND schemes referred to in Table 6.1 should receive more attention. The two-norm of tracking errors, i.e., $\| \mathbf{e}(t) \|_2$, is utilized to evaluate the performance of different models. Comparisons of tracking errors

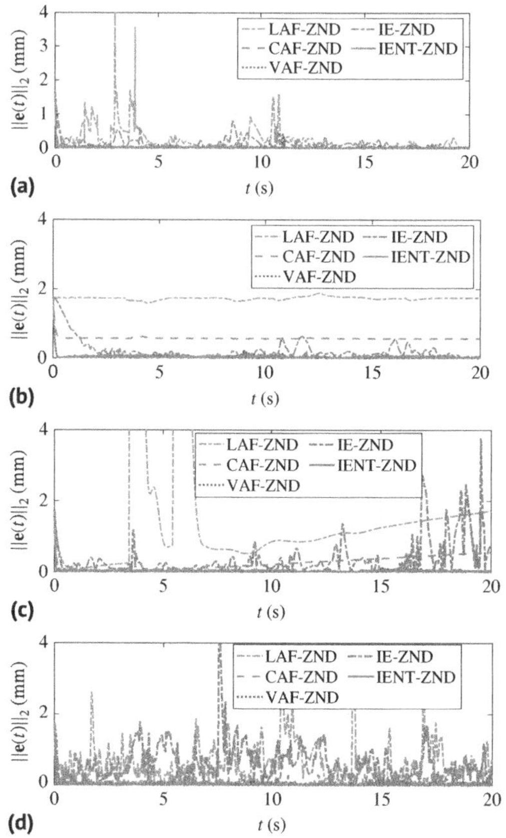

Figure 6.5 Comparisons on two norms of tracking errors (i.e., $\| \mathbf{e}(t) \|_2$) for continuum robot when employing different ZND models under disturbance of different types of noises: (a) noise-free; (b) constant noise $\eta(t) = 5$; (c) time-varying noise $\eta(t) = 5t/T_{\mathrm{d}}$; (d) bounded random noise.

(i.e., $\| \mathbf{e}(t) \|_2$) when employing different ZND models under the disturbance of different types of noises are depicted in Figure 6.5. When disturbed by noises, the ZND model can be modified to $\dot{e}(t) = -\beta\phi(e(t)) + \eta(t)$, where $\eta(t)$ denotes the noise. Firstly, as shown in Figure 6.5a, the dual ZND scheme achieves superior convergence and sufficiently small errors when employing these ZND models in the noise-free case, except for LAF-ZND and IE-ZND. When the control system is disturbed by constant noise ($\eta(t) = 5$), as illustrated in Figure 6.5b, both the VAF-ZND and IENT-ZND models exhibit noise suppression capability, as the errors almost converge to zero at $t = 0.21$ s. When the IE-ZND model is employed, it is evident that the noise can also be suppressed to some extent. However, both the LAF-ZND and CAF-ZND models demonstrate poor performance, as the errors do not show any tendency to converge. When the control system is disturbed by linear time-varying noise $\eta(t) = 5t/T_{\mathrm{d}}$, similar results are obtained, as illustrated in Figure 6.5c. Finally, random noise is also considered in the simulations, as shown in Figure 6.5d. It is readily observed that the IENT-ZND model also achieves the best performance compared with other ZND models. In this case, it is worth noting that the IE-ZND model demonstrates poor performance, which means this model is not suitable for cases involving random noise. In general, when employing the conventional LAF-ZND and CAF-ZND, the dual ZND control scheme has poor noise suppression capability. The IE-ZND is effective for suppressing constant noise and linear time-varying noise. However, it is not suitable for cases involving random noise. In comparison, it is observed that the tracking errors almost converge to zero in finite time under the disturbance of different noises when IENT-ZND model is adopted, thereby substantiating the robustness and efficacy of the IENT-ZND model and the control system. It is noteworthy that the VAF-ZND model performs similarly to the IENT-ZND model. To show the merits of the IENT-ZND model, the amplitude of the noise is increased ($\eta(t) = 10$), and the tracking error also does not exhibit a tendency to converge when the VAF-ZND model is adopted. In comparison, the tracking error still converges to zero gradually when the IENT-ZND model is adopted, due to the integration operation in the model. For a quantitative comparison, the RMSEs achieved by different models are summarized in Table 6.3.

The impact of parameter settings (i.e., the value of β) on the system requires further investigation. Theoretically, the value of β primarily influences the convergence time of the control system, as shown in Table 6.4. It is evident that the convergence time decreases as the parameter β increases when the control system is disturbed by different noises.

6.4.2 Experiments

In this part, experiments based on a cable-driven continuum robot are conducted to assess the effectiveness and noise-tolerant capability of the dual ZND control

Table 6.3 Comparisons on root-mean-square errors (RMSEs) synthesized by different control systems under disturbance of different noises, where unit of data is millimeter (mm).

Noise type	LAF	CAF	VAF	IE	IENT
Noise-free	0.197	0.126	0.123	0.178	0.122
Constant noise	1.357	0.437	0.125	0.280	0.124
Time-varying noise	0.975	0.315	0.123	0.194	0.123
Random noise	0.269	0.148	0.128	0.245	0.123

Table 6.4 Impact of parameter selection on convergence time.

β	Noise-free (s)	Constant noise (s)	Time-varying noise (s)	Random noise (s)
1	0.16	0.77	0.17	0.15
3	0.13	0.32	0.13	0.12
5	0.10	0.21	0.09	0.09

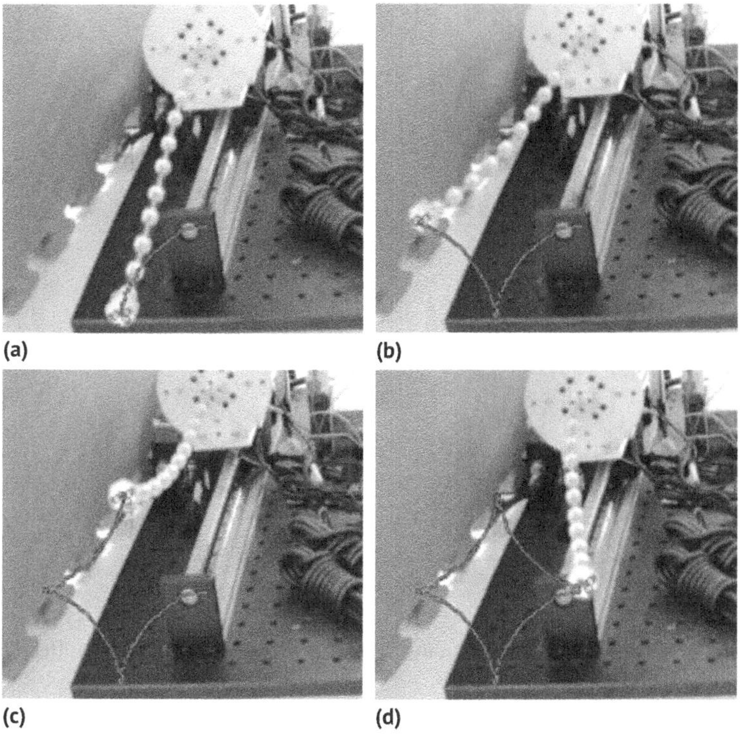

Figure 6.6 Some snapshots of continuum robot for tracking star path using dual ZND scheme: (a) $t = T_d/8$; (b) $t = 3T_d/8$; (c) $t = 5T_d/8$; (d) $t = T_d$.

system. The continuum robot is required to track a star path in the task space, with a duration of 150 s.

6.4.2.1 Experiment Results

Some snapshots of the continuum robot tracking the star path in the absence of noise are depicted in Figure 6.6, along with the actual trajectory displayed in the images. The specific experimental results achieved by the IENT-ZND model are illustrated in Figure 6.7. It is evident that the actual trajectory of the end-effector

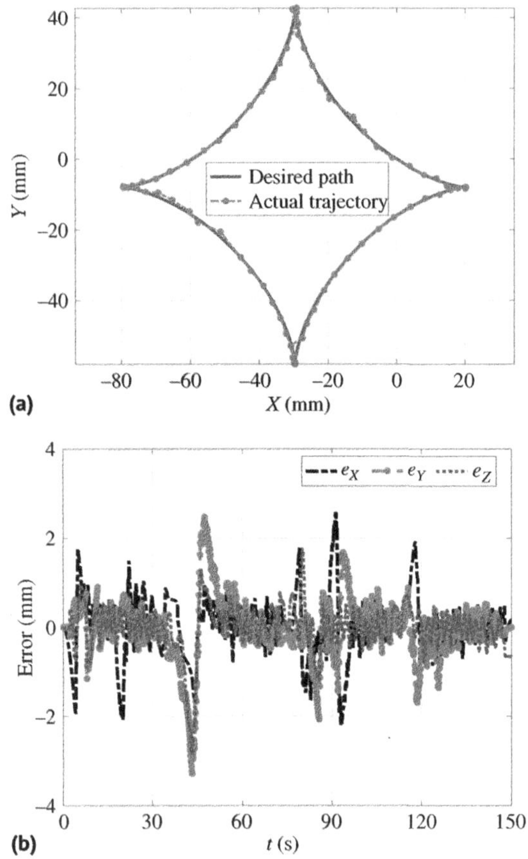

Figure 6.7 Experiment results of continuum robot for tracking star path: (a) end-effector trajectories; (b) tracking errors.

closely coincides with the desired path, and the RMSE of the tracking process is 1.12 mm. These experimental results imply that the IENT-ZND-based control system is feasible for the control of the continuum robot. It is worth noting that the tracking error cannot precisely converge to zero during experiments due to the imperfect controllability of the robot and the measurement error (0.2 mm for a static object) of the visual sensor.

Furthermore, the noise-tolerant capability of the control system should be validated in comparison with an existing model-free ZND-based (MFZND) method [106]. The experimental results are summarized in Figure 6.8, where the two norms of tracking errors ($\| \mathbf{e}(t) \|_2$) in different cases are illustrated. It is evident that, with the same parameter setting, the IENT-ZND-based control system

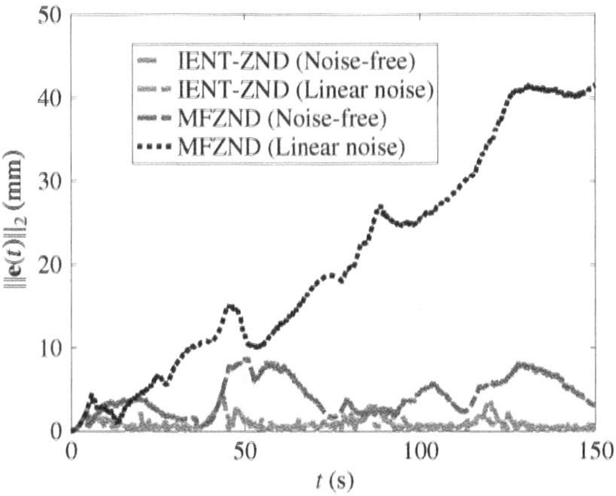

Figure 6.8 Comparisons on two norms of tracking errors (i.e., $\| \mathbf{e}(t) \|_2$) on physical continuum robot when employing different methods under disturbance of different types of noises.

achieves a lower tracking error than the MFZND method in the noise-free case, indicating the accuracy of the control system. Additionally, comparative experiments are performed in the presence of linear noise. It is evident that the MFZND method exhibits poor noise-tolerant capabilities, as the tracking error increases significantly when disturbed by noise. In comparison, the noise has a negligible influence on the IENT-ZND-based method, illustrating the superior noise-tolerant capabilities of the control system.

In summary, these experimental results have confirmed the effectiveness, accuracy and robustness of the scheme based on nonlinear ZND models for controlling continuum robots.

6.5 Chapter Summary

This chapter introduces several nonlinear ZND models, which improve the convergence and robustness of the linear ZND model. Based on these ZND models, a noise-tolerant dual ZND scheme is introduced for the tracking control of continuum robots. Comprising two nonlinear ZND models, the control system resolves the inverse kinematics problem with superior convergence and robustness. Finally, extensive simulations and experiments based on continuum robots have been conducted to validate the control method.

7

Varying-parameter Zeroing Neurodynamic Control Methods

7.1 Introduction

Traditional rigid robots are actuated by motors located at each joint, and the pose of the end-effector is determined from the given robot parameters and joint angles. However, their performance degrades in constrained and uncertain environments [14]. In contrast, continuum robots are made of flexible materials that allow for significant deformation [107], enabling more agile movement in confined environments [108]. They are thus well-suited for a wide range of applications, including wearable devices [109] and surgical assistance [110]. However, such soft structures and unconventional actuation methods introduce numerous nonlinearities, making accurate modeling of continuum robots highly challenging [111]. In recent years, researchers have made significant progress in the motion control of continuum robots. The explored control approaches can generally be categorized into two types: model-based methods and data-driven methods.

Model-based control methods simplify the physical structure of continuum robots and approximate their characteristics using analytical functions, thereby establishing a mapping between actuation space and task space. For example, Qiu et al. [112] employed a piecewise constant curvature (PCC) model to represent the continuum robot geometry as multiple arc segments with constant curvature and proposed an efficient multi-solution solver that resolves the inverse kinematics problem through iterative search. Thamo et al. [113] utilized the Cosserat rod theory to model the robot backbone as a series of infinitesimal rigid segments with infinite degrees of freedom and proposed a nonlinear solver for the Cosserat rod equations to rapidly estimate their solutions, thereby significantly improving computational efficiency. In contrast to the slender-body assumption required by Cosserat rod theory and the bending-specific modeling of the PCC model, the finite element model (FEM) discretizes the robot into a limited number of small elements. It does not require an explicit geometric model and is capable of handling large deformations under soft fluidic actuation. FEM can predict performance under various input conditions and aids researchers in understanding the influence of local strain on global behavior [114]. To reduce computational complexity, model-based control approaches often simplify robot dynamics and neglect nonlinear behaviors, which in turn compromises the model's accuracy.

Data-driven control methods utilize nonparametric modeling to learn the forward and inverse kinematics of continuum robots from large-scale training data. Typical techniques include Gaussian mixture models (GMMs) [115], Gaussian process regression (GPR) [116], feedforward neural networks (FNNs) [117], recursive neural networks (RNNs) [118], and reinforcement learning [119]. These approaches typically employ machine learning strategies to construct regression models that learn inverse kinematics from large datasets. However, their performance on redundant continuum robots remains suboptimal. Moreover, most existing data-driven methods require collecting large amounts of training data in advance, which can significantly reduce the operational lifespan of continuum robotic devices. Furthermore, applying these methods to different robotic platforms often requires recollecting data and retraining the models, which is both time-consuming and labor-intensive.

To avoid the complexity of analytical modeling for robots while enabling online learning, Zhang and Guo [27] proposed a novel recurrent neural network (RNN) model termed zeroing neurodynamics (ZND). The ZND is formulated based on differential equations, where variables, parameters, and weights are defined by the user, thereby eliminating the need for explicit training. It has found broad applications in robotic control. Early implementations mainly adopted fixed-parameter ZND (FP-ZND). For instance, Zhang et al. [120] applied FP-ZND to solve Sylvester equations with time-varying coefficient matrices, while [121] used them to address tracking control problems of parallel robots with unknown parameters. To enhance the control accuracy of ZND, researchers developed varying-parameter ZND (VP-ZND), where the parameters vary over time. Zhang and Yan [122] formulated the control of redundant robots as a time-varying convex quadratic programming problem and proposed the VP-ZND for its effective solution. To achieve accurate solutions in noisy environments, Zhang et al. [55], inspired by control theory, incorporated an integral term into the ZND formulation to solve the Moore–Penrose inverse of time-varying full-rank matrices. This method showed strong noise-suppression capability and was applied to robotic trajectory tracking.

The Jacobian matrix of a robot characterizes the relationship between actuation velocity and the Cartesian velocity of the end effector. Solving inverse kinematics problems typically requires computing the pseudo-inverse (i.e., Moore–Penrose inverse) of the Jacobian. Many existing control approaches adopt incremental methods, wherein each element of the robot's state parameter matrix is perturbed by a small increment, and the resulting displacement of the end effector is measured to sequentially compute each column of the Jacobian. The pseudo-inverse is then computed via singular value decomposition (SVD). However, as the Jacobian is often high-dimensional, directly computing the pseudo-inverse at every time step within a RNN is computationally expensive. One potential solution involves using two ZND models: one to solve the inverse kinematics and obtain the joint parameters and the other to estimate the inverse of the Jacobian matrix online, thereby avoiding the need to compute its pseudo-inverse at each time step.

This chapter is structured around the following key aspects. First, an inverse kinematics solution method based on VP-ZND models is presented to address the kinematic control problem in continuum robots. Second, a VP-ZND-based approach is introduced to directly adapt the pseudo-inverse of the Jacobian matrix, removing the computational overhead associated with calculating the pseudo-inverse. Additionally, two distinct VP-ZND models are investigated and compared to implement a model-free control strategy for soft continuum robots. Finally, simulations are conducted to validate the effectiveness and superiority of the VP-ZND control schemes.

Recursive least squares (RLS) is an online algorithm for estimating time-varying parameters. By employing a "forgetting" strategy, RLS assigns lower weights to older data and higher weights to more recent data, enabling real-time parameter updates. For example, Hu et al. [123] applied RLS to enable real-time online calibration of the state-of-charge parameters in lithium-ion batteries. This chapter also introduces a data-driven control framework based on VP-ZND and RLS [124]. The framework requires no pre-training and enables online estimation of model information for continuum robots, thereby achieving effective control of end-effector motion trajectories.

7.2 Preliminary

Generally, the kinematic model of a continuum robot is given by

$$\mathbf{p}(t) = \mathbf{f}(\mathbf{u}(t)), \tag{7.1}$$

where $\mathbf{p}(t) \in \mathbb{R}^m$ denotes the m-dimensional task-space state, $\mathbf{u}(t) \in \mathbb{R}^n$ is the n-dimensional actuation state, and $\mathbf{f} : \mathbb{R}^n \mapsto \mathbb{R}^m$ is the mapping function. Taking the first-order time derivative of the above model gives

$$\dot{\mathbf{p}}(t) = J(t)\dot{\mathbf{u}}(t), \tag{7.2}$$

where $J(t) = \partial \mathbf{f}(\mathbf{u}(t))/\partial \mathbf{u} \in \mathbb{R}^{m \times n}$ is the Jacobian matrix of the robot.

This chapter focuses on the trajectory tracking control problem of continuum robots, where the objective is to determine suitable actuation signals $\dot{\mathbf{u}}(t)$ or $\mathbf{u}(t)$ such that the robot's end effector follows a desired task-space path $\mathbf{p}_\mathrm{d} \in \mathbb{R}^m$. If an accurate and fully known model of the robot is available, the required control signals can be computed via the inverse mapping of the position-level kinematic equation (7.1) or the velocity-level kinematics of (7.2). However, in practice, due to the flexible nature of continuum robots and the influence of various factors, such as gravitational effects, external disturbances, and actuation coupling, the precise analytical modeling becomes extremely challenging, which in turn limits the applicability of traditional model-based control methods to continuum robots. To address this issue, this chapter introduces a model-free control framework for the motion control of continuum robots, capable of achieving accurate trajectory tracking without relying on explicit analytical models. For brevity, the argument t will be dropped hereinafter if no confusion is caused.

7.3 Varying-parameter ZND Methods

This section introduces two control methods based on VP-ZND.

7.3.1 Control Scheme Based on VP-ZND-1

Firstly, a vector-valued error function is defined as follows to monitor the process of solving the tracking control problem:

$$\mathbf{e} = \mathbf{p}_\mathrm{d} - \mathbf{p}, \tag{7.3}$$

where $\mathbf{p}_\mathrm{d} \in \mathbb{R}^m$ is the user-defined path that the end effector needs to follow. Then, the goal is to solve the inverse kinematics problem to obtain appropriate control signal $\mathbf{u}$ that forces $\mathbf{e}$ to converge to zero. This section adopts the following VP-ZND-1 evolution law [122]:

$$\frac{\mathrm{d}\mathbf{e}}{\mathrm{d}t} = -\gamma \exp(t)\mathbf{g}(\mathbf{e}), \tag{7.4}$$

where $\gamma > 0$ is a constant scalar designed to scale the convergence rate, and $\mathbf{g}(\cdot) : \mathbb{R}^m \to \mathbb{R}^m$ is the activation function processing array. The linear activation function is applied in this chapter (i.e., $\mathbf{g}(\mathbf{e}) = \mathbf{e}$). Substituting (7.3) into (7.4) leads to

$$\dot{\mathbf{p}}_\mathrm{d} - \dot{\mathbf{p}} = -\gamma \exp(t)(\mathbf{p}_\mathrm{d} - \mathbf{p}), \tag{7.5}$$

where $\dot{\mathbf{p}}_\mathrm{d}$ and $\dot{\mathbf{p}}$ are derivatives of $\mathbf{p}_\mathrm{d}$ and $\mathbf{p}$ with respect to time t, respectively. Substituting the velocity-level robot model into (7.5) yields

$$\dot{\mathbf{p}}_\mathrm{d} - J\dot{\mathbf{u}} = -\gamma \exp(t)(\mathbf{p}_\mathrm{d} - \mathbf{p}), \tag{7.6}$$

where $\mathbf{u} \in \mathbb{R}^n$ is the variable to be solved. For (7.6), the solution is as follows:

$$\dot{\mathbf{u}} = J^{\dagger}(\dot{\mathbf{p}}_{\rm d} + \gamma \exp(t)(\mathbf{p}_{\rm d} - \mathbf{p})), \tag{7.7}$$

where $J^{\dagger} \in \mathbb{R}^{n\times m}$ is the pseudo-inverse of J. For (7.7), $\mathbf{p}_{\rm d}$ and $\dot{\mathbf{p}}_{\rm d}$ are defined by users, and $\mathbf{p}$ is obtained by sensor equipment. To solve the differential equation and get the actuation signal $\mathbf{u}$, we still need to determine the value of $J^{\dagger}$. In model-free control schemes, the model of robots is regarded as unknown. Therefore, the Jacobian matrix J can not be calculated directly based on its definition. The common method in existing works is to estimate the Jacobian matrix of soft continuum robots and to calculate its pseudo-inverse. Inspired by Chen et al. [121], this chapter directly estimates the pseudo-inverse of the Jacobian matrix in order to eliminate the cost of calculating the pseudo-inverse. Replacing $J^{\dagger}$ with X, one can rewrite (7.7) as the following neurodynamic form:

$$\dot{\mathbf{u}} = X(\dot{\mathbf{p}}_{\rm d} + \gamma \exp(t)(\mathbf{p}_{\rm d} - \mathbf{p})). \tag{7.8}$$

The next step is to estimate X with another VP-ZND-1 model.

Based on the velocity-level kinematic model of the continuum robot, we have

$$\dot{\mathbf{u}} = J^{\dagger}\dot{\mathbf{p}}. \tag{7.9}$$

Similarly, a vector-valued error function is defined to monitor the process of estimating X:

$$\epsilon = \dot{\mathbf{u}} - X\dot{\mathbf{p}}. \tag{7.10}$$

Utilizing the VP-ZND-1 evolution law (7.4) again, one gets

$$\frac{{\rm d}\epsilon}{{\rm d}t} = \ddot{\mathbf{u}} - \dot{X}\dot{\mathbf{p}} - X\ddot{\mathbf{p}} = -\gamma \exp(t)(\dot{\mathbf{u}} - X\dot{\mathbf{p}}), \tag{7.11}$$

where $\ddot{\mathbf{u}}$ and $\ddot{\mathbf{p}}$ are the second derivatives of $\mathbf{u}$ and $\mathbf{p}$ with respect to t, respectively, and $\dot{X}$ is the derivative of X. Based on (7.11), the following neurodynamic equation is obtained:

$$\dot{X} = (\ddot{\mathbf{u}} - X\ddot{\mathbf{p}} + \gamma \exp(t)(\dot{\mathbf{u}} - X\dot{\mathbf{p}}))\dot{\mathbf{p}}^{\dagger}, \tag{7.12}$$

where $\dot{\mathbf{p}}^{\dagger}$ is the pseudo-inverse of $\dot{\mathbf{p}}$. What needs to be clarified is that the velocity $\dot{\mathbf{p}}$ and acceleration $\ddot{\mathbf{p}}$ of the end effector are obtained by sensory equipment. Besides, $\ddot{\mathbf{u}}$ is approximated by Euler backward formula:

$$\ddot{\mathbf{u}} = \frac{\dot{\mathbf{u}}(t) - \dot{\mathbf{u}}(t - \Delta t)}{\Delta t}, \tag{7.13}$$

where Δt is selected in the light of accuracy requirements. With the given initial value, the neurodynamic equations (7.8) and (7.12) are solved to obtain the required control signal $\mathbf{u}$, which is used to steer the continuum robot.

7.3.2 Control Scheme Based on VP-ZND-2

The VP-ZND-1 model was designed to improve the convergence of the original ZND model. However, it is unable to guarantee the finite-time convergence, which is more meaningful in actual applications. VP-ZND-2 [47] is able to make up for this shortcoming and has the capability of suppressing noise. For a predefined error function $\boldsymbol{\varepsilon} \in \mathbb{R}^m$, VP-ZND-2 has the following evolution law:

$$\frac{\mathrm{d}\varepsilon}{\mathrm{d}t} = -\gamma \mathbf{q}(\varepsilon) - \gamma \int_0^t \mathbf{h}(\varepsilon(\tau))\mathrm{d}\tau, \tag{7.14}$$

where γ is a constant scalar, and $\mathbf{q}(\cdot) : \mathbb{R}^m \to \mathbb{R}^m$ and $\mathbf{h}(\cdot) : \mathbb{R}^m \to \mathbb{R}^m$ denote two activation function arrays. The entries of $\mathbf{q}(\cdot)$ and $\mathbf{h}(\cdot)$ adopted in the VP-ZND-2 model are defined as follows, respectively:

$$\begin{cases} q(x) &= \left(k_1\sigma^{\frac{1}{2}}(t,x)|x|^{\frac{1}{2}} + k_2\sigma(t,x)|x|\right)\mathrm{sign}(x) \\ h(x) &= \left(k_3\sigma(t,x) + k_4\sigma^2(t,x)|x|\right)\mathrm{sign}(x), \end{cases} \tag{7.15}$$

where parameter k_i with $i = 1, 2, 3, 4$ is a positive constant and variable gain $\sigma(x, t)$ satisfying $\dot{\sigma}(x, t) = c\mathrm{sign}(|x|)$ with constant $c > 0$ has a positive initial state.

Similar to the last part, substituting the error functions (7.3) and (7.10) into the evolution law (7.14), respectively, leads to the following neurodynamic equations:

$$\dot{\mathbf{u}} = X\left(\dot{\mathbf{p}}_\mathrm{d} + \gamma\mathbf{q}(\mathbf{e}) + \gamma\int_0^t \mathbf{h}(\mathbf{e}(\tau))\mathrm{d}\tau\right), \tag{7.16}$$

$$\dot{X} = \left(\ddot{\mathbf{u}} - X\ddot{\mathbf{p}} + \gamma\mathbf{q}(\epsilon) + \gamma\int_0^t \mathbf{h}(\epsilon(\tau))\mathrm{d}\tau\right)\dot{\mathbf{p}}^\dagger. \tag{7.17}$$

With given initial values, the above neurodynamic equations are solved in a similar way to estimate the pseudo-inverse of Jacobian matrix and obtain the required control signal $\mathbf{u}$. Compared with the control scheme (i.e., (7.8) and (7.12)) based on the VP-ZND-1 model, the control scheme (i.e., (7.16) and (7.17)) based on the VP-ZND-2 model is able to achieve finite-time convergence and has noise-suppression capability benefited from the integral operation.

For (7.14), if the activation function satisfies $q(x) = \exp(t)x$ and $h(x) = 0$, it is evident that the control scheme based on the VP-ZND-2 model (i.e., (7.16) and (7.17)) will degenerate to the control scheme based on the VP-ZND-1 model (i.e., (7.8) and (7.12)). Finally, the schematic diagram of the control scheme is distinctly depicted in Figure 7.1. The inputs of the control system are user-defined path $\mathbf{p}_\mathrm{d}$ and desired velocity $\dot{\mathbf{p}}_\mathrm{d}$. With the user-defined inputs, the estimated

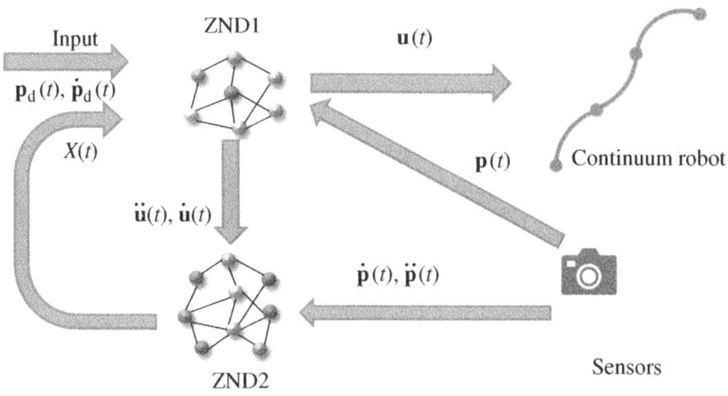

Figure 7.1 Schematic diagram of control scheme.

pseudo-inverse X of Jacobian matrix and the actual trajectory $\mathbf{p}$ of the end effector measured by sensors, the inverse kinematics is solved by ZND1 to attain the required control signal $\mathbf{u}$. At the same time, the pseudo-inverse adaption of the time-varying Jacobian matrix is implemented by ZND2. The control scheme guarantees better convergence and robustness by means of VP-ZND models.

7.4 Recursive Least Square-aided Varying-parameter ZND Method

To achieve trajectory tracking for continuum robots, the corresponding state parameters $\mathbf{u}$ are generated from the desired Cartesian path $\mathbf{p}_\mathrm{d}$, such that $\mathbf{f}(\mathbf{u}) = \mathbf{p} \to \mathbf{p}_\mathrm{d}$. The control scheme presented in this section employs ZND to solve the inverse kinematics problem. Furthermore, the RLS method is utilized to update the pseudo-inverse of the Jacobian matrix estimate, thereby avoiding the time-consuming task of directly computing the Jacobian matrix and its pseudo-inverse in each iteration. Additionally, the closed-loop control mechanism ensures that the control error is minimized.

7.4.1 Inverse Kinematics Based on VP-ZND

Firstly, to record the real-time trajectory tracking process of the robot, the error between the actual trajectory of the robot's end effector and the desired path is defined as the following vector-valued error function [125]:

$$\mathbf{e} = \mathbf{p}_\mathrm{d} - \mathbf{p}. \tag{7.18}$$

To ensure that all elements e_i (where $i = 1, 2, \dots, m$) of the vector $\mathbf{e}$ converge to zero, a ZND design formula, as shown in Zhang et al. [21], is employed:

$$\frac{\mathrm{d}\mathbf{e}}{\mathrm{d}t} = -\lambda \boldsymbol{\phi}(\mathbf{e}), \tag{7.19}$$

where $\lambda > 0$ is the convergence coefficient. When λ is a constant, the model is referred to as a FP-ZND. When λ is time-varying, the model is called a VP-ZND. $\boldsymbol{\phi}(\cdot) : \mathbb{R}^m \mapsto \mathbb{R}^m$ is the activation function vector to be defined. Substituting (7.18) into (7.19), we obtain the following expression:

$$\dot{\mathbf{p}} = \dot{\mathbf{p}}_{\mathrm{d}} + \lambda \boldsymbol{\phi}(\mathbf{e}), \tag{7.20}$$

where $\dot{\mathbf{p}}$ and $\dot{\mathbf{p}}_{\mathrm{d}}$ are the first-order time derivatives of $\mathbf{p}$ and $\mathbf{p}_{\mathrm{d}}$ with respect to time t, respectively. Substituting (7.20) into (7.2), we obtain

$$\dot{\mathbf{u}} = J^{\dagger} \left(\dot{\mathbf{p}}_{\mathrm{d}} + \lambda \boldsymbol{\phi}(\mathbf{e})\right). \tag{7.21}$$

Given a desired path $\mathbf{p}_{\mathrm{d}}$ and its corresponding velocity $\dot{\mathbf{p}}_{\mathrm{d}}$, together with the end-effector position $\mathbf{p}$ obtained from sensors, (7.21) can be solved to determine the corresponding actuation state $\mathbf{u}$ that drives the continuum robot to track the desired path.

Scholars have conducted extensive research on the design formulations of ZND, revealing significant differences among various models in terms of stability and convergence. For instance, the original ZND [27] adopts a simple formulation but suffers from low control accuracy. The design parameters of VP-ZND [122] increase exponentially. However, its convergence time remains infinite, and it fails to guarantee that the steady-state error converges to zero under additional noise disturbances. Tan et al. [47] and Tan et al. [126] introduced integral terms in their formulations, enabling finite-time convergence and improved noise robustness. Xiao et al. [127] extended VP-ZND to the complex domain for solving time-varying complex-valued inverse matrix problems. The underlying idea is that the design parameters should be relatively large when the error is significant and reduced accordingly as the error diminishes. This approach ensures fast convergence while avoiding unnecessarily large parameters.

This chapter presents an adapted dynamic varying-parameter enhanced ZND (ADVPEZND), whose design formula is

$$\frac{\mathrm{d}\mathbf{e}}{\mathrm{d}t} = -\lambda \exp\left(\xi_1 \beta^t \|\mathbf{e}\|_2\right) \mathbf{g}(\mathbf{e}) + \lambda \int_0^t \exp\left(\xi_2 \tau + \xi_3\right) \mathbf{q}(\mathbf{e}) \mathrm{d}\tau, \tag{7.22}$$

where $\xi_i > 0$ ($i = 1, 2, 3$), $\beta > 1$, and $\mathbf{g}(\cdot)$ and $\mathbf{q}(\cdot)$ are activation function arrays. The ith elements of the activation functions are defined as

$$\begin{cases} g_i(e_i) = \begin{cases} \zeta_1 \psi^{r_1}(e_i) + \zeta_2 e_i \exp(\zeta_3 |e_i| + 1), & |e_i| \leq 1 \\ \zeta_1 \psi^{r_2}(e_i) + \zeta_2 e_i \exp(\zeta_3 |e_i| + 1), & \text{otherwise} \end{cases} \\ q_i(e_i) = \begin{cases} \zeta_4 \psi^{r_1}(e_i), & |e_i| \leq 1 \\ \zeta_4 \psi^{r_2}(e_i), & \text{otherwise,} \end{cases} \end{cases} \tag{7.23}$$

where $\zeta_i > 0\,(i = 1, 2, 3, 4)$, $0 < r_1 < 1$, $r_2 > 1$ are design parameters, and $\psi^{r_1}(e_i)$ and $\psi^{r_2}(e_i)$ are defined as

$$\psi^{\mu}(e_i) = \begin{cases} e_i^{\mu}, & e_i > 0 \\ 0, & e_i = 0 \\ -|e_i|^{\mu}, & e_i < 0 \end{cases}. \tag{7.24}$$

Substituting (7.18) into (7.22) yields

$$\dot{\mathbf{p}} = \dot{\mathbf{p}}_{\mathrm{d}} + \lambda \exp(\xi_1 \beta^t \parallel \mathbf{e} \parallel_2) \mathbf{g}(\mathbf{e}) + \lambda \int_0^t \exp(\xi_2 \tau + \xi_3) \mathbf{q}(\mathbf{e}) \mathrm{d}\tau. \tag{7.25}$$

Considering the velocity-level kinematics, we have

$$\dot{\mathbf{u}} = J^{\dagger} \left(\dot{\mathbf{p}}_{\mathrm{d}} + \lambda \exp(\xi_1 \beta^t \parallel \mathbf{e} \parallel_2) \mathbf{g}(\mathbf{e}) + \lambda \int_0^t \exp(\xi_2 \tau + \xi_3) \mathbf{q}(\mathbf{e}) \mathrm{d}\tau \right). \tag{7.26}$$

The design parameters of the ADVPEZND do not grow unbounded over time, ensuring long-term model stability. In addition, the inclusion of error integral terms enables finite-time convergence and effective noise suppression.

7.4.2 Jacobian Estimation Based on RLS

Since the nonlinear mapping $\mathbf{f}(\cdot)$ between the robot's state parameter $\mathbf{u}$ and the end-effector position $\mathbf{p}$ in (7.1) is unknown, the Jacobian matrix J cannot be computed analytically via its definition. Therefore, an estimation method for J must be developed. It is important to note that the inverse kinematics formulation involves the pseudo-inverse of the Jacobian matrix, $J^{\dagger}$. Computing the pseudo-inverse at every time step is computationally expensive. To address this, the RLS algorithm is employed to directly estimate $J^{\dagger}$. According to (7.2), the following relation holds:

$$\dot{\mathbf{u}}^{\mathrm{T}} = \dot{\mathbf{p}}^{\mathrm{T}} J^{\dagger^{\mathrm{T}}}. \tag{7.27}$$

Let $X = J^{\dagger^{\mathrm{T}}}$, we have $J^{\dagger} = X^{\mathrm{T}}$.

First, we define $\omega = \dot{\mathbf{u}}^{\mathrm{T}} - \dot{\mathbf{p}}^{\mathrm{T}}\hat{X}$, where $\hat{X}$ denotes the estimation of X. To get $\hat{X}$, the following optimization problem is designed:

$$\begin{aligned} \min y &= \omega^{\mathrm{T}}\omega = \sum_{i=1}^{n} \omega_i^2 \\ &= \left(\dot{\mathbf{u}}^{\mathrm{T}} - \dot{\mathbf{p}}^{\mathrm{T}}\hat{X}\right)^{\mathrm{T}} \left(\dot{\mathbf{u}}^{\mathrm{T}} - \dot{\mathbf{p}}^{\mathrm{T}}\hat{X}\right) \\ &= \dot{\mathbf{u}}\dot{\mathbf{u}}^{\mathrm{T}} - \hat{X}^{\mathrm{T}}\dot{\mathbf{p}}\dot{\mathbf{u}}^{\mathrm{T}} - \dot{\mathbf{u}}\dot{\mathbf{p}}^{\mathrm{T}}\hat{X} + \hat{X}^{\mathrm{T}}\dot{\mathbf{p}}\dot{\mathbf{p}}^{\mathrm{T}}\hat{X} \end{aligned} \tag{7.28}$$

$$\text{s. t.} \quad \dot{\mathbf{u}} = \hat{X}^{\mathrm{T}}\left(\dot{\mathbf{p}} + \lambda\boldsymbol{\phi}(\mathbf{e})\right).$$

The minimum of the cost function y is achieved at its extremum point. By setting $\frac{\partial y}{\partial \hat{X}} = 0$, the following solution is obtained:

$$\hat{X} = \left(\dot{\mathbf{p}}\dot{\mathbf{p}}^{\mathrm{T}}\right)^{-1}\dot{\mathbf{p}}\dot{\mathbf{u}}^{\mathrm{T}}. \tag{7.29}$$

It is noted that the evolution of $\hat{X}$ is a continuous process, and the estimate at time step k, $\hat{X}_k$, may depend on the results obtained in the previous time steps, such as $\hat{X}_{k-1}, \cdots, \hat{X}_1$. Furthermore, leveraging the previous estimates to predict $\hat{X}_k$ can yield a result that better aligns with the behavior of the continuum robot model itself. Therefore, (7.29) is extended as follows:

$$\hat{X}_k = \left(\dot{\mathcal{P}}_k\dot{\mathcal{P}}_k^{\mathrm{T}}\right)^{-1}\dot{\mathcal{P}}_k\dot{U}_k^{\mathrm{T}}, \tag{7.30}$$

where

$$\dot{\mathcal{P}}_k = \left[\sqrt{\gamma}\dot{\mathcal{P}}_{k-1} \quad \dot{\mathbf{p}}_k\right], \tag{7.31}$$

$$\dot{U}_k = \left[\sqrt{\gamma}\dot{U}_{k-1} \quad \dot{\mathbf{u}}_k\right], \tag{7.32}$$

with $\gamma \in (0, 1]$ being a forgetting factor. Then, we have

$$\begin{aligned} \dot{\mathcal{P}}_k\dot{\mathcal{P}}_k^{\mathrm{T}} &= \left[\sqrt{\gamma}\dot{\mathcal{P}}_{k-1} \quad \dot{\mathbf{p}}_k\right]\begin{bmatrix} \sqrt{\gamma}\dot{\mathcal{P}}_{k-1}^{\mathrm{T}} \\ \dot{\mathbf{p}}_k^{\mathrm{T}} \end{bmatrix} \\ &= \gamma\dot{\mathcal{P}}_{k-1}\dot{\mathcal{P}}_{k-1}^{\mathrm{T}} + \dot{\mathbf{p}}_k\dot{\mathbf{p}}_k^{\mathrm{T}} \end{aligned} \tag{7.33}$$

and

$$\begin{aligned} \dot{\mathcal{P}}_k\dot{U}_k^{\mathrm{T}} &= \left[\sqrt{\gamma}\dot{\mathcal{P}}_{k-1} \quad \dot{\mathbf{p}}_k\right]\begin{bmatrix} \sqrt{\gamma}\dot{U}_{k-1}^{\mathrm{T}} \\ \dot{\mathbf{u}}_k^{\mathrm{T}} \end{bmatrix} \\ &= \gamma\dot{\mathcal{P}}_{k-1}\dot{U}_{k-1}^{\mathrm{T}} + \dot{\mathbf{p}}_k\dot{\mathbf{u}}_k^{\mathrm{T}}. \end{aligned} \tag{7.34}$$

By substituting (7.33) and (7.34) into (7.29) and defining $K_k = (\dot{\mathcal{P}}_k\dot{\mathcal{P}}_k^{\mathrm{T}})^{-1}$, the following expression is obtained:

$$
\begin{aligned}
\hat{\mathcal{X}}_k &= (\dot{\mathcal{P}}_k\dot{\mathcal{P}}_k^{\mathrm{T}})^{-1}\dot{\mathcal{P}}_k\dot{\mathcal{U}}_k^{\mathrm{T}} \\
&= (\dot{\mathcal{P}}_k\dot{\mathcal{P}}_k^{\mathrm{T}})^{-1}\left(\gamma\dot{\mathcal{P}}_{k-1}\dot{\mathbf{U}}_{k-1}^{\mathrm{T}} + \dot{\mathbf{p}}_k\dot{\mathbf{u}}_k^{\mathrm{T}}\right) \\
&= (\dot{\mathcal{P}}_k\dot{\mathcal{P}}_k^{\mathrm{T}})^{-1}\left((\dot{\mathcal{P}}_k\dot{\mathcal{P}}_k^{\mathrm{T}} - \dot{\mathbf{p}}_k\dot{\mathbf{p}}_k^{\mathrm{T}})\hat{\mathcal{X}}_{k-1} + \dot{\mathbf{p}}_k\dot{\mathbf{u}}_k^{\mathrm{T}}\right) \\
&= \hat{\mathcal{X}}_{k-1} + K_k\dot{\mathbf{p}}_k\left(\dot{\mathbf{u}}_k^{\mathrm{T}} - \dot{\mathbf{p}}_k^{\mathrm{T}}\hat{\mathcal{X}}_{k-1}\right).
\end{aligned}
\tag{7.35}
$$

The above expression indicates that the current estimate $\hat{\mathcal{X}}_k$ at time step k can be recursively computed from the previous estimate $\hat{\mathcal{X}}_{k-1}$. Although (7.29) still involves the matrix inversion of $\dot{\mathcal{P}}_k\dot{\mathcal{P}}_k^{\mathrm{T}}$, it can be reformulated using the Sherman–Morrison–Woodbury formula [128] as follows:

$$
\begin{aligned}
K_k &= (\dot{\mathcal{P}}_k\dot{\mathcal{P}}_k^{\mathrm{T}})^{-1} \\
&= \left(\gamma\dot{\mathcal{P}}_{k-1}\dot{\mathcal{P}}_{k-1}^{\mathrm{T}} + \dot{\mathbf{p}}_k\dot{\mathbf{p}}_k^{\mathrm{T}}\right)^{-1} \\
&= (\gamma\dot{\mathcal{P}}_{k-1}\dot{\mathcal{P}}_{k-1}^{\mathrm{T}})^{-1} - (\gamma\dot{\mathcal{P}}_{k-1}\dot{\mathcal{P}}_{k-1}^{\mathrm{T}})^{-1}\times \\
&\quad\ \dot{\mathbf{p}}_k\left(I + \dot{\mathbf{p}}_k^{\mathrm{T}}(\gamma\dot{\mathcal{P}}_{k-1}\dot{\mathcal{P}}_{k-1}^{\mathrm{T}})^{-1}\dot{\mathbf{p}}_k\right)^{-1}\dot{\mathbf{p}}_k^{\mathrm{T}}(\gamma\dot{\mathcal{P}}_{k-1}\dot{\mathcal{P}}_{k-1}^{\mathrm{T}})^{-1} \\
&= \frac{1}{\gamma}\left(K_{k-1} - \frac{K_{k-1}\dot{\mathbf{p}}_k\dot{\mathbf{p}}_k^{\mathrm{T}}K_{k-1}}{\gamma + \dot{\mathbf{p}}_k^{\mathrm{T}}K_{k-1}\dot{\mathbf{p}}_k}\right).
\end{aligned}
\tag{7.36}
$$

By combining (7.35) and (7.36), the estimated value $\hat{\mathcal{X}}$ of $J^{\dagger}$ can be computed, which is then incorporated into the inverse kinematics formulation (i.e., (7.26)) to achieve trajectory tracking for the continuum robot.

7.5 Verification of VP-ZND-1 and VP-ZND-2

To verify the feasibility and reveal the merits of the control schemes, simulations are carried out based on a two-segment continuum robot in this section. The simulations are performed as follows.

7.5.1 Settings

To begin with, there are some parameters and variables that need to be clarified in the control schemes. The constant scalar is set to be $\gamma = 1$. The initial state of the actuation signal is $\mathbf{u}(0) = [10, 0, 0, -10, 0, 0,]^{\mathrm{T}}$ mm. The parameters of the

VP-ZND-2 model are always set as $k_1 = k_2 = 1$, $k_3 = 2$, and $k_4 = 5$. The duration of the tracking task is $T_\mathrm{d} = 4$ s. Besides, it is necessary to estimate the initial value of Jacobian matrix of the continuum robot. By independently increasing the ith actuation signal (i.e., $u_i(0)$) by an incremental amount Δu, the displacement $\Delta \mathbf{p}_i$ of the end effector can be measured. Next, the ith column of the Jacobian matrix is initialized as $J_i(0) = \Delta \mathbf{p}_i / \Delta u$, and $X(t)$ is initialized as $X(0) = J^\dagger(0)$. In addition, the model parameters of the soft continuum robot also need to be specified for the sake of simulating the soft continuum robot. The distance from the center of the segment to those arcs is set as $d = 20$ mm. The initial backbone length of the ith segment is set to be $l_{i0} = 100$ mm for the two-segment robot.

7.5.2 Simulation Results

In this section, simulations are conducted based on the two-segment soft continuum robot. The robot is required to track a three-ring path in the three-dimensional task space.

The motion results synthesized by the control scheme based on the VP-ZND-2 model are depicted in Figure 7.2. As shown in Figure 7.2a, the actual trajectory of the end effector almost coincides with the desired path. Figure 7.2b records the tracking errors in terms of different coordinate axes. One can find that the errors finally converge to zero in finite time after experiencing a period of fluctuation. Besides, Figure 7.2c depicts the changes of the control signal during the task. These results have verified the feasibility of the control scheme for continuum robots.

Likewise, simulations have also been performed when adopting the control schemes based on the VP-ZND-1 model and the original ZND (O-ZND) model [106]. Figure 7.3 illustrates the comparative results of the simulations. As shown in Figure 7.3a, VP-ZND models can reduce the tracking error of soft continuum robots significantly compared with the O-ZND model in the noise-free case. Besides, when disturbed by noise, the performance of the control schemes is also worth studying. In the noise-polluted control system, the neurodynamic equation (7.8) is modified as

$$\dot{\mathbf{u}} = X(\dot{\mathbf{p}}_\mathrm{d} + \gamma \exp(t)(\mathbf{p}_\mathrm{d} - \mathbf{p}) + \boldsymbol{\eta}) \tag{7.37}$$

and the neurodynamic equation (7.16) is modified as

$$\dot{\mathbf{u}} = X(t)\left(\dot{\mathbf{p}}_\mathrm{d} + \gamma \mathbf{q}(\mathbf{e}) + \gamma \int_0^t \mathbf{h}(\mathbf{e}(\tau))\mathrm{d}\tau + \boldsymbol{\eta}\right) \tag{7.38}$$

where $\boldsymbol{\eta}$ is the additional noise. When disturbed by linear time-varying noise $\eta_i(t) = 30t/T_\mathrm{d}$, there are residual tracking errors when employing the O-ZND and VP-ZND-1 models, as depicted in Figure 7.3b. However, the tracking error almost converges to zero when adopting the VP-ZND-2 model, which reveals the superior robustness of the control method based on the VP-ZND-2 model.

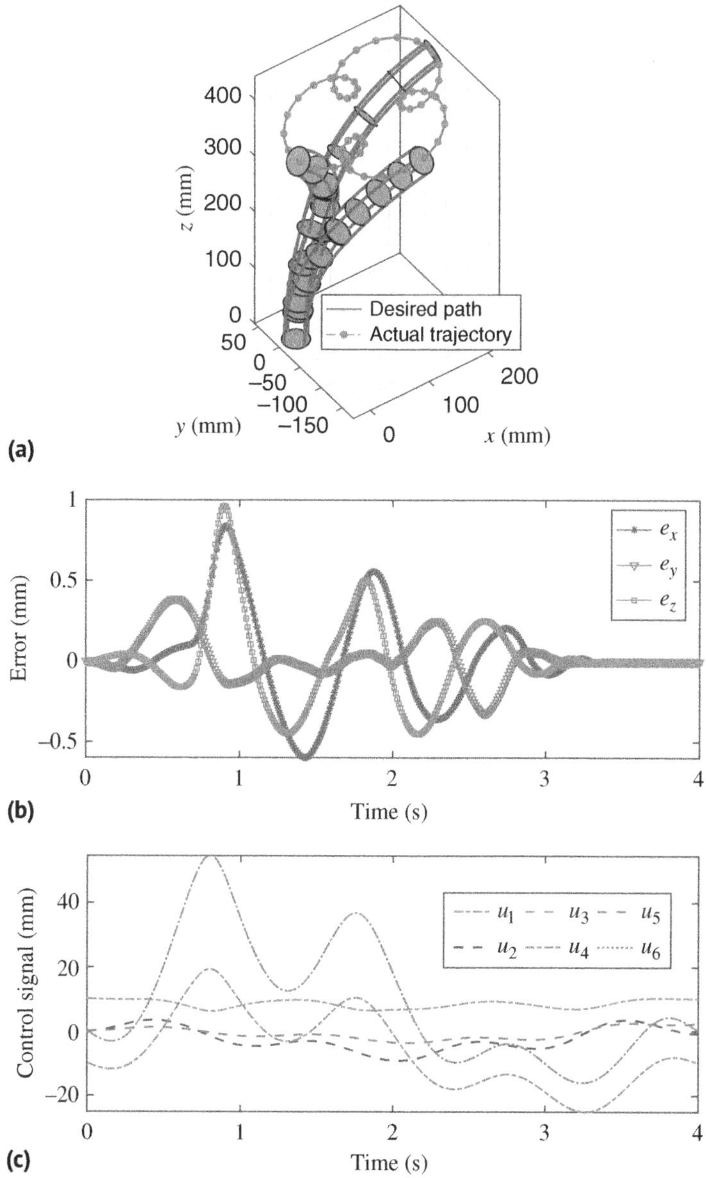

Figure 7.2 Three-ring path tracking results of two-segment soft continuum robot when adopting control scheme based on VP-ZND-2: (a) motion of continuum robot; (b) tracking errors; (c) control signal.

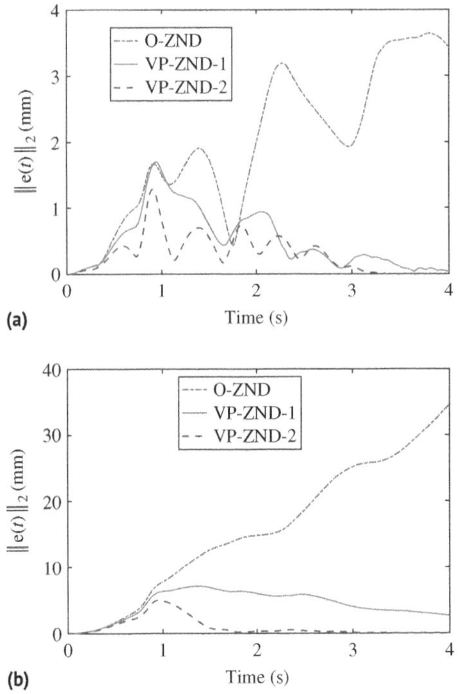

Figure 7.3 Tracking errors of two-segment soft continuum robot when adopting control schemes based on different ZND models: (a) with $\eta_i(t) = 0$; (b) with $\eta_i(t) = \dfrac{30t}{T_{\mathrm{d}}}$.

7.6 Verification of ADVPEZND

This section provides simulation results to validate the effectiveness of the ADVPEZND method for trajectory tracking in continuum robots.

7.6.1 Setting

The parameter settings for the control scheme are first specified, with a total task duration of $T_{\mathrm{d}} = 4$ s. The convergence coefficient is set to $\lambda = 50$. The initial state of a three-segment continuum robot is $\mathbf{u}(0) = [12, 0, 0, -12, 0, 0, 12, 0, 0]^{\mathrm{T}}$ mm. The parameters of the ADVPEZND model are empirically optimized through multiple rounds of simulations and are set as follows: $\beta = 2$, $\xi_1 = 10$, $\xi_2 = 1$, $\xi_3 = 5$, $\zeta_1 = 1$, $\zeta_2 = 1$, $\zeta_3 = 1$, $\zeta_4 = 0.1$, $r_1 = 0.8$, and $r_2 = 3$.

In (7.35), the matrix K is initialized as the identity matrix. The matrix $\hat{X}$ is initialized using an incremental method, where the ith column of the Jacobian matrix J is estimated by perturbing the ith element of $\mathbf{u}(0)$ by a small increment

Δu and measuring the resulting displacement $\Delta \mathbf{p}_i$ of the end effector, such that $J_i(0) = \Delta \mathbf{p}_i / \Delta u$. The initial estimate is then obtained as $\hat{X}(0) = J^{\dagger}(0)$.

For equations involving integral terms, numerical solutions are obtained using the explicit Euler method. A small discrete time step $\Delta t = 10^{-4}$ s is used to ensure slow variation between consecutive time steps and maintain the system's stability. The task space of the continuum robot is defined in three-dimensional Cartesian space, and the desired trajectory is given by $\mathbf{p}_\mathrm{d} = \mathbf{p}_\mathrm{a} + \mathbf{p}_\mathrm{d}(0)$, where $\mathbf{p}_\mathrm{a}$ is defined as follows:

$$\mathbf{p}_\mathrm{a} = \alpha \begin{bmatrix} \cos\left(4\pi \sin^2\left(\frac{\pi t}{2T_\mathrm{d}}\right)\right) \cos\left(2\pi \sin^2\left(\frac{\pi t}{2T_\mathrm{d}}\right)\right) - 1 \\ \frac{\sqrt{3}}{2} \cos\left(4\pi \sin^2\left(\frac{\pi t}{2T_\mathrm{d}}\right)\right) \sin\left(2\pi \sin^2\left(\frac{\pi t}{2T_\mathrm{d}}\right)\right) \\ \frac{1}{2} \cos\left(4\pi \sin^2\left(\frac{\pi t}{2T_\mathrm{d}}\right)\right) \sin\left(2\pi \sin^2\left(\frac{\pi t}{2T_\mathrm{d}}\right)\right) \end{bmatrix}, \tag{7.39}$$

where a scaling factor $\alpha = 120$ mm is applied to adjust the scale of the desired path, ensuring that the actuation signal $\mathbf{u}$ of the continuum robot remains within a reasonable range. The term $\mathbf{p}_\mathrm{d}(0)$ represents the initial position of the end effector.

7.6.2 Results

The simulation results based on the control scheme using the ADVPEZND model are illustrated in Figure 7.4. The actual trajectory of the continuum robot's end effector closely follows the desired path. Figure 7.5a depicts the tracking errors

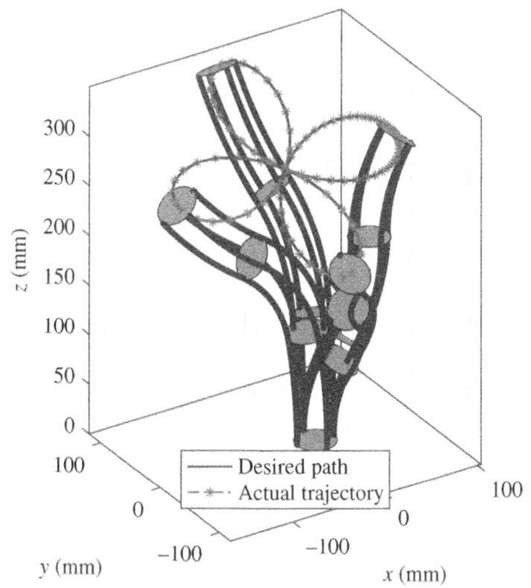

Figure 7.4 Trajectory tracking of three-segment continuum robot using ADVPEZND.

along the three Cartesian axes. Although slight fluctuations are observed during the tracking process, the tracking errors eventually converge to zero. The sharp fluctuations at certain moments are primarily caused by overshooting in the integral term of the control formulation, especially at trajectory turning points. Nevertheless, the maximum tracking error remains below 0.1 mm, which is acceptable for the task. The root-mean-square error (RMSE) throughout the tracking task is 0.057 mm. Figure 7.5b records the evolution of the actuation signal $\mathbf{u}$. These simulation results confirm the effectiveness and feasibility of the control scheme.

In the control system, λ plays a critical role. Table 7.1 illustrates its impact on the tracking error of the ADVPEZND controller. As the convergence factor λ increases, the tracking error is significantly reduced. However, if λ becomes too large, it amplifies noise and may induce overshooting, potentially leading to trajectory divergence.

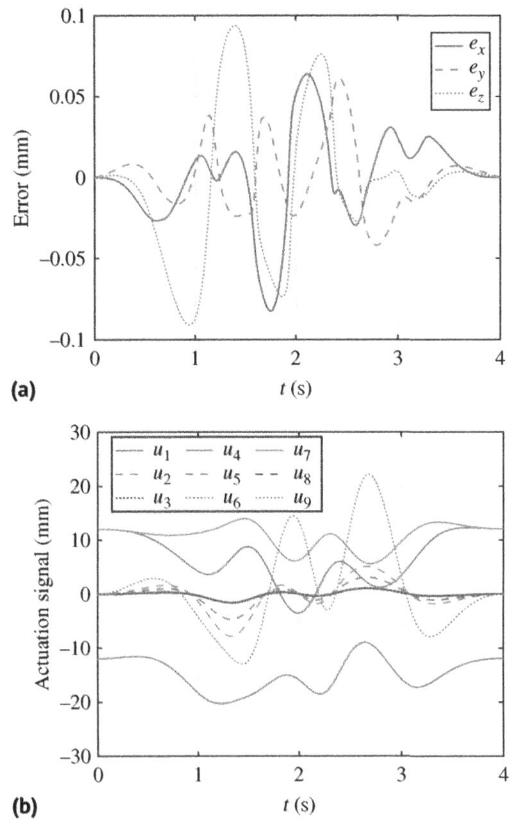

Figure 7.5 Tracking results in simulation: (a) tracking errors $\mathbf{e} = [e_x, e_y, e_z]^{\mathrm{T}}$ along different axes; (b) changes of actuation signal $\mathbf{u}$.

Table 7.1 Quantitative analysis of convergence factor λ.

λ	1	3	5	10	30	50
RMSE (mm)	0.150	0.122	0.110	0.093	0.068	0.057

7.7 Chapter Summary

This chapter addresses the tracking control problem of soft continuum robots. Two model-free control schemes based on distinct VP-ZND models (i.e., VP-ZND-1 and VP-ZND-2) are introduced. Notably, these schemes circumvent the need to calculate the pseudo-inverse of the Jacobian matrix. More importantly, when utilizing the VP-ZND-2 model, the control strategy ensures finite-time convergence of the tracking error while effectively suppressing noise. In addition, a control scheme combining VP-ZND and RLS is introduced to address the trajectory tracking problem of continuum robots. The improved ADVPEZND model is integrated into the control framework, utilizing a ZND model to solve the inverse kinematics and achieve precise end-effector trajectory control. Notably, the approach introduces the RLS method into time-varying robot control problems, enabling the online estimation of the pseudo-inverse of the Jacobian matrix. This eliminates the reliance on an explicit analytical model of the continuum robot and avoids the computational burden of matrix inversion. Simulations have been conducted to validate the feasibility and robustness of these control schemes.

8

Prescribed-performance Convergent Zeroing Neurodynamic Control Methods

8.1 Introduction

Over the past few decades, numerous modeling methods have been investigated as foundations for designing model-based control methods. Despite significant progress in model-based control of continuum robots, it is argued that mathematical modeling of continuum robots remains challenging due to incomplete sensing of their configurations and susceptibility to environmental interactions [14]. These concerns have led to the development of model-free control methods for continuum robots [19, 51], eliminating the need for both analytical and numerical models. However, numerous unexplored challenges remain in the field of continuum robot control. This chapter focuses on two primary challenges. First, previous control methods for continuum robots often struggle to achieve desired task-space performance, such as convergence time, transient error, and residual error. This has typically been accomplished by laboriously fine-tuning control system parameters based on engineering experience or extensive experimental testing. Developing a control system that can achieve desired task-space performance without parameter fine-tuning remains a significant knowledge gap in continuum robot control. Second, the diversity of continuum robots presents substantial challenges for developing a generic control system. Model-based and learning-based methods typically require establishing or learning different models for various robots, which is labor-intensive. Although adaptive methods potentially offer rapid adaptation to different robots, there is a lack of studies exploring and validating this capability.

Generally, continuum robots are required to reach a reference point or follow a reference trajectory, often starting with an initial error in motion control tasks. As the robot end-effector approaches the reference point or trajectory, the associated task-space error is termed transient error, which is followed by steady-state error. The primary objective of this chapter is to achieve predefined behavior performance in the kinematic control of continuum robots. The predefined behavior encompasses two aspects: prescribed task-space error and predefined convergence time. This issue is critical because robots are typically required to complete tasks with prescribed accuracy within a predefined time. The prescribed performance control (PPC) theory, which enables users to directly prescribe the convergence rate of transient error and error boundaries, has been extensively investigated [129], including applications to the configuration control of continuum robots [130, 131]. However, research on task-space PPC for continuum robots has been barely reported. Moreover, most existing PPC methods only function in ideal conditions without external disturbances, while various factors can cause errors to exceed prescribed boundaries, ultimately leading to task failure. Particularly, continuum robots are more susceptible to disturbances due to their softness, presenting significant challenges for their PPC.

To meet the demands of various scenarios, a variety of heterogeneous continuum robots (HCRs) have been designed [132–134]. The emergence of HCRs has subsequently led to the demand for developing robot-specific or task-specific control methods. However, this is laborious and requires significant effort due to the diversity of continuum robots. Therefore, the second objective of this chapter is to develop a generic control method for HCRs. Generally, these HCRs differ primarily in overall length (ranging from 50 to 500 mm), radial diameter (ranging from 1 to 50 mm), degrees of freedom (DoF) in actuation space (ranging from 2 to 6 DoF), actuation mechanism (including motors and pneumatic muscles), and task space (including 3D Cartesian space and 2D image plane). These differences significantly degrade the efficiency of existing model-based or data-driven control strategies when applied to different physical setups, as it requires substantial time and effort to build or learn robot-specific models. Variations in other aspects of these robots also influence existing control strategies. For instance, external forces and gravity significantly influence the configurations of cable-driven continuum robots (CCRs) but have less effect on other setups.

This chapter presents a unified motion control framework for HCRs, leveraging advances in zeroing neurodynamics (ZND) to achieve predefined-time convergence and synchronized motion control. The chapter is motivated by the need for precise and time-constrained performance. We begin by introducing a novel control paradigm: the predefined-time convergent ZND (PTC-ZND) model. Theoretical analysis rigorously proves that this model guarantees convergence of the error function to zero within a user-defined time, independent of initial conditions. Leveraging two instances of the PTC-ZND model, we develop an inverse kinematics solver and a state estimator tailored for continuum robots, enabling predefined-time motion control. The effectiveness of this method is demonstrated on two distinct types of continuum robots: a CCR and a concentric-tube robot (CTR).

To address the limitations of conventional ZND models, particularly their inability to ensure synchronized convergence, we further present a novel predefined-time synchronized ZND (PTS-ZND) framework. Unlike existing predefined-time stable systems [135], PTS control ensures that all state variables converge simultaneously within the predefined time, significantly improving trajectory efficiency, convergence accuracy, and energy usage [136]. We provide the first theoretical analysis showing that conventional ZND models lack this PTS property and develop a new PTS-ZND formulation to overcome this limitation. This framework is extended to support trajectory tracking, and an iterative gradient dynamics-based Jacobian learning method is incorporated to enable model-free control. Finally, we integrate PPC theory with ZND to design a behavior-prescribed controller that guarantees user-defined transient and steady-state performance without the need for manual parameter tuning [137]. To handle the challenge of model dependency, we present an adaptive control scheme that combines the controller with an online ZND-based estimator of the velocity-level kinematic model. The control architecture is extensively validated on five types of HCRs, including CCR, CTR, parallel continuum robots (PCRs), flexible endoscopes, and pneumatic continuum robots (PneumCRs), demonstrating its generality, robustness, and practical applicability.

8.2 Predefined-time Convergent Method

This section presents the PTC-ZND model [138] and proves its predefined-time convergence theoretically. Subsequently, this section introduces how to design a state estimator and an inverse kinematics solver for continuum robots with the PTC-ZND model.

8.2.1 PTC-ZND Model

In the design procedure of ZND, an indefinite error-monitoring function $\xi(t) \in \mathbb{R}$ is first defined at time t according to the specific objective. To drive the error-monitoring function to converge to zero, the design formula of ZND is usually given as follows:

$$\dot{\xi}(t) = -\nu\phi\,(\xi(t)), \tag{8.1}$$

where $\dot{\xi}(t)$ denotes the first time derivative of $\xi(t)$, ν is a positive constant dominating the convergence speed of the model, and $\phi\,(\cdot)$ could be a linear or nonlinear activation function. To achieve predefined-time convergence, the following PTC-ZND model is introduced:

$$\begin{cases} \dot{\xi}(t) = -\nu\phi\left(\xi(t)\right) \\ \phi(z) = \dfrac{\exp\left(|z|^{2-2r} + |z|^{2+2r}\right)|z|^{2r-1}\mathrm{sign}(z)}{2\left(1 - r + (1+r)|z|^{4r}\right)} \\ \nu = \dfrac{1 - \exp\left(-|\xi(0)|^{2-2r} - |\xi(0)|^{2+2r}\right)}{t_c} \end{cases}, \tag{8.2}$$

where $r \in (0.5, 1)$ is a parameter of the model, sign(·) is the sign function, $\xi(0)$ denotes the initial value of the error-monitoring function, and t_c is the desired convergence time defined by users. The choice of r can ensure that the activation function $\phi(z)$ is monotonically increasing provided that the input z is in a reasonable range. Finally, by expanding the ZND model with user-defined error-monitoring function $\xi(t)$, one can obtain a differential-equation-based dynamic system to solve specific problems. Regarding the PTC-ZND model, the following theorem is provided to ensure its predefined-time convergence.

Theorem 8.1 *Starting from initial error $\xi(0) \neq 0$, the PTC-ZND model* (8.2) *ensures that the error-monitoring function $\xi(t)$ converges to zero in predefined time t_c.*

Proof. To begin with, the following Lyapunov candidate function is defined:

$$l(t) = \xi^2(t) \geq 0. \tag{8.3}$$

By deriving the above function with respect to time t, the following result is attained:

$$\dot{l}(t) = 2\xi(t)\dot{\xi}(t). \tag{8.4}$$

By utilizing (8.2) and replacing the variable z in $\phi(z)$ with ξ, we further obtain

$$\begin{aligned} \dot{l}(t) &= -2\nu\xi(t)\phi\left(\xi(t)\right) \\ &= \frac{-\nu|\xi(t)|\exp\left(|\xi(t)|^{2-2r} + |\xi(t)|^{2+2r}\right)|\xi(t)|^{2r-1}}{1 - r + (1+r)|\xi(t)|^{4r}} \\ &= \frac{-\nu\exp\left(|\xi(t)|^{2-2r} + |\xi(t)|^{2+2r}\right)}{(1-r)|\xi(t)|^{-2r} + (1+r)|\xi(t)|^{2r}}. \end{aligned} \tag{8.5}$$

The above equation is then rewritten as

$$\dot{l}(t) = \frac{\mathrm{d}l}{\mathrm{d}t} = \frac{-\nu\exp\left(l^{1-r}(t) + l^{1+r}(t)\right)}{(1-r)l^{-r}(t) + (1+r)l^{r}(t)}, \tag{8.6}$$

which yields

$$-\frac{(1-r)l^{-r}(t)+(1+p)l^{r}(t)}{\exp\left(l^{1-r}(t)+l^{1+r}(t)\right)}\mathrm{d}l=\nu \mathrm{d}t. \tag{8.7}$$

It is supposed that the error-monitoring function $\xi(t)$ converges to zero at the time t_s, which means the function $l(t)$ converges to zero at t_s. By integrating both sides of (8.7) from $t = 0$ to t_s, the following result is obtained:

$$\begin{aligned}
\int_{l(0)}^{0} -\frac{(1-r)l^{-r}(t)+(1+r)l^{r}(t)}{\exp\left(l^{1-r}(t)+l^{1+r}(t)\right)}\mathrm{d}l &= \int_{0}^{t_s} \nu \mathrm{d}t, \\
\exp\left(-l^{1-r}(t)-l^{1+r}(t)\right)\Big|_{l(0)}^{0} &= \nu t\Big|_{0}^{t_s}, \\
1-\exp\left(-l^{1-r}(0)-l^{1+r}(0)\right) &= \nu t_s.
\end{aligned} \tag{8.8}$$

Considering the definition of ν in (8.2), we have

$$\begin{aligned}
t_s &= \frac{1-\exp\left(-l^{1-r}(0)-l^{1+r}(0)\right)}{\nu} \\
&= \frac{1-\exp\left(-l^{1-r}(0)-l^{1+r}(0)\right)}{1-\exp\left(-|\xi(0)|^{2-2r}-|\xi(0)|^{2+2r}\right)}t_c \\
&= t_c.
\end{aligned} \tag{8.9}$$

The above result indicates that the convergence time of the error-monitoring function $\xi(t)$ is the predefined time t_c. The proof is thus completed.

8.2.2 Jacobian Estimator

For model-based control methods, there are two major problems: 1) The modeling of continuum robots is of great difficulty and complexity. 2) It is tedious to design specific models and control algorithms for HCRs. Therefore, we are aimed at developing a generic model-free control system, which is independent of the robot model, to achieve the motion control of HCRs.

Typically, the forward kinematics of a continuum robot is simply described by

$$\mathbf{p}(t)=\mathbf{f}(\mathbf{u}(t)), \tag{8.10}$$

where $\mathbf{p}(t) \in \mathbb{R}^m$ denotes the position of robot end-effector in the m-dimensional task space, $\mathbf{u}(t) \in \mathbb{R}^n$ denotes the state of the n actuators of the continuum robot, and $\mathbf{f}(\cdot)$ denotes a nonlinear kinematic mapping function. Accordingly, the differential kinematics of the continuum robot is as follows:

$$\dot{\mathbf{p}}(t)=J(t)\dot{\mathbf{u}}(t), \tag{8.11}$$

where $\dot{\mathbf{p}}(t)$ denotes the first time derivative of $\mathbf{p}(t)$ and $J(t) = \partial\mathbf{f}(\mathbf{u}(t))/\partial\mathbf{u}(t) \in \mathbb{R}^{m\times n}$ is the Jacobian matrix of the continuum robot. Without building a model for the robot, the mapping function and the Jacobian matrix cannot be analytically calculated. To solve this problem, a Jacobian-based state estimator is designed for the continuum robot.

First, inspired by (8.11), the following vector-valued error-monitoring function is defined:

$$\epsilon(t) = \dot{\mathbf{p}}(t) - J_{\mathrm{A}}(t)\dot{\mathbf{u}}(t) \in \mathbb{R}^m, \tag{8.12}$$

where $J_{\mathrm{A}}(t)$ is the unknown Jacobian matrix to be estimated for the continuum robot. Then, modifying the PTC-ZND model to vector-valued version and replacing $\xi(t)$ with $\epsilon(t)$ lead to

$$\dot{\epsilon}(t) = -\nu_{\mathrm{s}}\boldsymbol{\phi}(\epsilon(t)), \tag{8.13}$$

where $\boldsymbol{\phi}(\cdot) \in \mathbb{R}^m$ is an activation function array with each element being $\phi(z)$ and ν_{s} is a constant designed according to (8.2). Next, by expanding (8.13) with (8.12), the following dynamic equation is derived:

$$\ddot{\mathbf{p}}(t) - \dot{J}_{\mathrm{A}}(t)\dot{\mathbf{u}}(t) - J_{\mathrm{A}}(t)\ddot{\mathbf{u}}(t) = -\nu_{\mathrm{s}}\boldsymbol{\phi}(\dot{\mathbf{p}}(t) - J_{\mathrm{A}}(t)\dot{\mathbf{u}}(t)), \tag{8.14}$$

where $\ddot{\mathbf{p}}(t)$ denotes the second time derivative of $\mathbf{p}(t)$. Finally, the evolution of the estimated Jacobian matrix follows the dynamic equation:

$$\dot{J}_{\mathrm{A}}(t) = (\ddot{\mathbf{p}}(t) - J_{\mathrm{A}}(t)\ddot{\mathbf{u}}(t) + \nu_{\mathrm{s}}\boldsymbol{\phi}(\dot{\mathbf{p}}(t) - J_{\mathrm{A}}(t)\dot{\mathbf{u}}(t)))\dot{\mathbf{u}}^{\dagger}(t), \tag{8.15}$$

where $\dot{\mathbf{u}}^{\dagger}(t)$ denotes the pseudo-inverse of $\dot{\mathbf{u}}(t)$.

8.2.3 Predefined-time Convergent Inverse Kinematics Solver

Similarly, to achieve the predefined-time convergent motion control of the continuum robot, the following vector-valued error-monitoring function is defined:

$$\mathbf{e}(t) = \mathbf{p}_{\mathrm{d}}(t) - \mathbf{p}(t) \in \mathbb{R}^m. \tag{8.16}$$

In a tracking control task, $\mathbf{p}_{\mathrm{d}}(t)$ denotes the reference path that needs to be tracked by robot end-effector. In a point-to-point task, we can simply set $\mathbf{p}_{\mathrm{d}}(t)$ as the reference point, which implies the point-to-point task is a special case of the tracking control task. The objective is to drive the error-monitoring function $\mathbf{e}(t)$ to zero in predefined time. By adopting the PTC-ZND model (8.2) again, we have

$$\dot{\mathbf{e}}(t) = -\nu_{\mathrm{c}}\boldsymbol{\phi}(\mathbf{e}(t)), \tag{8.17}$$

where ν_{c} is a constant designed according to (8.2). Then, expanding (8.17) with (8.16) yields

$$\dot{\mathbf{p}}_{\mathrm{d}}(t) - J(t)\dot{\mathbf{u}}(t) = -\nu_{\mathrm{c}}\boldsymbol{\phi}(\mathbf{p}_{\mathrm{d}}(t) - \mathbf{p}(t)). \tag{8.18}$$

Therefore, the actuation signal can be determined by the following dynamic equation:

$$\dot{\mathbf{u}}(t) = J^{\dagger}(t)\,(\dot{\mathbf{p}}_{\mathrm{d}}(t) + \nu_{\mathrm{c}}\boldsymbol{\phi}(\mathbf{p}_{\mathrm{d}}(t) - \mathbf{p}(t)))\,. \tag{8.19}$$

However, the analytic Jacobian $J(t)$ cannot be directly used because the kinematic model of the continuum robot is unknown. By replacing the analytic Jacobian $J(t)$ with the estimated Jacobian $J_{\mathrm{A}}(t)$, the adaptation of the actuation signal follows the dynamic equation:

$$\dot{\mathbf{u}}(t) = J_{\mathrm{A}}^{\dagger}(t)\,(\dot{\mathbf{p}}_{\mathrm{d}}(t) + \nu_{\mathrm{c}}\boldsymbol{\phi}(\mathbf{p}_{\mathrm{d}}(t) - \mathbf{p}(t)))\,. \tag{8.20}$$

8.2.4 Control System for HCRs

Finally, a control system based on the state estimator (8.15) and the inverse kinematics solver (8.20) is obtained for the model-free motion control of continuum robots. There are several facts worth pointing out regarding the control system. First, taking advantage of the PTC-ZND model, the control system can drive the tracking error (in tracking tasks) or position error (in point-to-point tasks) to zero in predefined time. Second, it is not difficult to find that the control method is independent of the analytic model and structure of continuum robots, which means it is applicable to HCRs. Third, the selection of the convergence speed parameter in the PTC-ZND model is a key step to ensure the convergence of the error-monitoring function. Generally, the convergence speed ν is set according to (8.2). Specifically, in the inverse kinematics solver (8.20), the convergence speed is set as

$$\nu_{\mathrm{c}} = \frac{1 - \exp\left(-|e_{\max}(0)|^{2-2r} - |e_{\max}(0)|^{2+2r}\right)}{t_{\mathrm{c}}}, \tag{8.21}$$

where $e_{\max}(0)$ denotes the element in $\mathbf{e}(0)$ with the largest absolute value. In the state estimator (8.15), the convergence speed parameter is set to satisfy $\nu_{\mathrm{s}} \geq 1/t_{\mathrm{c}}$ for simplicity. Besides, it is observed from the dynamic equations (8.15) and (8.20) that the end-effector position, velocity, and acceleration are required to solve the equations. Unfortunately, the velocity and acceleration are not easy to be directly measured sometimes, which could be a limitation of the method. In this case, one can utilize differentiators or filters to estimate the velocity and acceleration. Moreover, the dynamic equations (8.15) and (8.20) could be solved by numerical tools (e.g., ode45 in MATLAB [139]) as long as the initial values are provided. The initial value of the actuation signal could be set to any reasonable value, whereas the initial value of the estimated Jacobian should follow specific procedures. Generally, one can increase the value of the ith $(i = 1, 2, \dots, n)$ actuation signal by a sufficiently small amount $\Delta u_i \in \mathbb{R}^+$ and measure the end-effector displacement $\Delta\mathbf{p}_i \in \mathbb{R}^m$ of the continuum robot. By repeating the procedure for each actuator, one can initialize the estimated Jacobian as $J_{\mathrm{A}}(0) = [\Delta\mathbf{p}_1/\Delta u_1, \Delta\mathbf{p}_2/\Delta u_2, \dots, \Delta\mathbf{p}_n/\Delta u_n]$.

8.3 PTS Method

This section presents a PTS control method for continuum robots.

8.3.1 Preliminary

To begin with, some basic knowledge is presented to facilitate the understanding of the following content.

8.3.1.1 Zeroing Neurodynamics

For a time-variant system with its state being $\mathbf{x}(t) = [x_1(t), \dots, x_n(t)]^{\mathrm{T}} \in \mathbb{R}^n$, to drive the system state to the equilibrium point where $\mathbf{x}(t) = \mathbf{0}$, the system dynamics is formulated as the following vector-valued ZND model:

$$\dot{\mathbf{x}}(t) = -\xi \mathbf{g}(\mathbf{x}(t)), \tag{8.22}$$

where $\dot{\mathbf{x}}(t)$ indicates the first time derivative of $\mathbf{x}(t)$, $\xi > 0$ controls the convergence rate of the system, and

$$\mathbf{g}(\mathbf{x}(t)) = [g(x_1(t)), \cdots, g(x_n(t))]^{\mathrm{T}} \in \mathbb{R}^n \tag{8.23}$$

is an activation function array with g being any monotonically increasing odd function. Generally, ZND takes the following steps to deal with time-variant problems: (1) Defining a proper error-monitoring function as the system state. (2) Substituting the error-monitoring function into the ZND model (8.22) to expand the model. (3) Calculating the required input to the system based on the expanded model. To accelerate the convergence of the linear ZND (L-ZND) model, nonlinear activation functions have been proposed, where the vector-valued sign function

$$\mathbf{sgn}_{\mathrm{c}}(\mathbf{x}) = [\mathrm{sgn}(x_1), \cdots, \mathrm{sgn}(x_n)]^{\mathrm{T}} \tag{8.24}$$

is commonly used. Specially, the result regarding the PTC-ZND model is presented below.

Theorem 8.2 (See Li [135]): *The system state* $\mathbf{x}(t)$ *of the ZND model* (8.22) *converges to the equilibrium point in predefined time* t_{p} *if the following activation function is adopted:*

$$g(x_i) = \left(|x_i|^{1-\frac{\rho}{2}} + |x_i|^{1+\frac{\rho}{2}}\right) \mathrm{sgn}(x_i), \tag{8.25}$$

where $\rho \in (0, 1)$ *is a parameter.*

8.3.1.2 Time-synchronized Control

To achieve TSC, the following norm-normalized sign function is commonly used:

$$\mathbf{sgn}_{\mathrm{n}}(\mathbf{x}) = \begin{cases} \dfrac{\mathbf{x}}{\|\mathbf{x}\|_2}, & \mathbf{x} \neq \mathbf{0} \\ \mathbf{0}, & \mathbf{x} = \mathbf{0} \end{cases}, \tag{8.26}$$

where $\|\mathbf{x}\|_2$ denotes the L_2-norm of $\mathbf{x}$. Then, some results regarding TSC are introduced below.

Definition 8.1 (See Li et al. [140]: *Ratio Persistence*): The state $\mathbf{x}(t)$ of a dynamic system is ratio persistent if

$$\frac{\mathbf{x}(t)}{\|\mathbf{x}(t)\|_2} = \zeta \frac{\dot{\mathbf{x}}(t)}{\|\dot{\mathbf{x}}(t)\|_2} \tag{8.27}$$

for $\mathbf{x}(t) \neq 0$, where ζ is the direction of ratio persistence, which could be either -1 or 1.

Definition 8.2 (See Li et al. [140]: *Time-Synchronized Stability*): The equilibrium of a dynamic system is time-synchronized stable if: (i) it is finite-time stable; (ii) for an open neighborhood $\mathcal{N}$ of the origin, there exists a synchronized settling-time function $T : \mathcal{N} \setminus \{0\} \to (0, \infty)$, such that $\forall \mathbf{x}(0) \in \mathcal{N} \setminus \{0\}$, $t \in (0, \infty)$, we have $\mathbf{x}(t) \in \mathcal{N}$, and for $x_i(0) \neq 0$, $\forall t \in [0, T(\mathbf{x}(0)))$, we have

$$x_i(t) \neq 0, \quad \lim_{t \to T(\mathbf{x}(0))} x_i(t) = 0. \tag{8.28}$$

According to Li et al. [140], a sufficient condition of time-synchronized stability is that the system state is ratio-persistent and the system is finite-time stable. Subsequently, the following definition regarding PTS stability is provided.

Definition 8.3 *(PTS Stability)*: The equilibrium of a dynamic system is PTS stable if: (i) its system state is ratio-persistent; (ii) its equilibrium is predefined-time stable.

8.3.2 PTS-ZND Model

This section analyzes the theoretical properties of different ZND models in terms of time-synchronized stability.

8.3.2.1 Conventional L-ZND and PTC-ZND Models

This part proves that conventional ZND models (i.e., L-ZND and PTC-ZND) are not PTS stable.

Theorem 8.3 *When adopting the linear activation function* $g(x_i(t)) = x_i(t)$, *the system state* $\mathbf{x}(t)$ *of the ZND (L-ZND) model* (8.22) *is ratio-persistent, but its equilibrium is not time-synchronized stable and not PTS stable.*

Proof: Based on the design formula (8.22), we have

$$\frac{\dot{\mathbf{x}}(t)}{\|\dot{\mathbf{x}}(t)\|_2} = \frac{-\xi\mathbf{g}(\mathbf{x}(t))}{\|\xi\mathbf{g}(\mathbf{x}(t))\|_2} = -\frac{\mathbf{x}(t)}{\|\mathbf{x}(t)\|_2}, \tag{8.29}$$

where $\mathbf{x}(t) \neq \mathbf{0}$. According to Definition 8.1 and (8.29), we can conclude that the system state of the L-ZND model is ratio-persistent. To analyze the stability and convergence, a Lyapunov candidate is defined as

$$l(t) = \mathbf{x}^{\mathrm{T}}(t)\mathbf{x}(t) \geq 0. \tag{8.30}$$

Taking the first time derivative of the candidate yields

$$\begin{aligned} \dot{l}(t) &= 2\mathbf{x}^{\mathrm{T}}(t)\dot{\mathbf{x}}(t) \\ &= -2\xi\mathbf{x}^{\mathrm{T}}(t)\mathbf{x}(t) \\ &= -2\xi l(t), \end{aligned} \tag{8.31}$$

which leads to

$$l(t) = l(0)\exp(-2\xi t). \tag{8.32}$$

Therefore, one gets $\lim_{t\to\infty} l(t) = 0$ because $l(0) = \mathbf{x}^{\mathrm{T}}(0)\mathbf{x}(0) \geq 0$ and $\xi > 0$. Although the L-ZND model is stable according to the Lyapunov's second method, it takes infinitely long time for the candidate to converge to zero, which means the L-ZND model is not finite-time stable and not predefined-time stable. Therefore, the first condition in Definition 8.2 is not satisfied, which means the L-ZND model is not time-synchronized stable. In addition, the second condition in Definition 8.3 is not satisfied, which means the L-ZND model is not PTS stable.

Theorem 8.4 *When adopting the activation function* (8.25), *the equilibrium of the ZND (PTC-ZND) model* (8.22) *is predefined-time stable, but its system state* $\mathbf{x}(t)$ *is not ratio-persistent, which means the PTC-ZND model is not PTS stable.*

Proof: Since it has been proven that the equilibrium of the PTC-ZND model is predefined-time stable [135], the relevant proof is omitted here. Next, considering the activation function (8.25) and the ZND model (8.22), we have

$$\begin{aligned} \frac{\|\dot{\mathbf{x}}(t)\|_2}{\|\mathbf{x}(t)\|_2}\mathbf{x}(t) &= \frac{\xi\|\mathbf{g}(\mathbf{x}(t))\|_2}{\|\mathbf{x}(t)\|_2}\mathbf{x}(t) \\ &= \Lambda_\delta\mathbf{x}(t), \end{aligned} \tag{8.33}$$

where Λ_δ is a diagonal matrix with

$$\delta(t) = \frac{\xi \|\mathbf{g}(\mathbf{x}(t))\|_2}{\|\mathbf{x}(t)\|_2} \tag{8.34}$$

being the diagonal elements. Besides, we also have

$$\begin{aligned} \dot{\mathbf{x}}(t) &= -\xi \mathbf{g}(\mathbf{x}(t)) \\ &= \Lambda_d \mathbf{x}(t), \end{aligned} \tag{8.35}$$

where Λ_d is a diagonal matrix with the ith diagonal element being

$$d_i(t) = -\xi \left(|x_i|^{-\rho/2} + |x_i|^{\rho/2}\right). \tag{8.36}$$

It is evident that $\exists \mathbf{x}(t) \neq \mathbf{0}$, such that $\Lambda_\delta \neq \zeta \Lambda_d$. Therefore, $\exists \mathbf{x}(t) \neq \mathbf{0}$, we have

$$\Lambda_\delta \mathbf{x}(t) \neq \zeta \Lambda_d \mathbf{x}(t), \tag{8.37}$$

which means

$$\frac{\mathbf{x}(t)}{\|\mathbf{x}(t)\|_2} \neq \zeta \frac{\dot{\mathbf{x}}(t)}{\|\dot{\mathbf{x}}(t)\|_2}. \tag{8.38}$$

Finally, according to Definition 8.1, we can conclude that the system state $\mathbf{x}(t)$ of the PTC-ZND model is not ratio-persistent. According to Definition 8.3, the PTC-ZND model is not PTS stable.

8.3.2.2 PTS-ZND Model

This part presents the following activation function for the ZND model (8.22):

$$\mathbf{g}(\mathbf{x}(t)) = \mathbf{sig}_{\mathrm{n}}^{1+\rho}(\mathbf{x}(t)) + \mathbf{sig}_{\mathrm{n}}^{1-\rho}(\mathbf{x}(t)), \tag{8.39}$$

where

$$\mathbf{sig}_{\mathrm{n}}^{\alpha}(\mathbf{x}(t)) = \|\mathbf{x}(t)\|_2^{\alpha} \mathbf{sgn}_{\mathrm{n}}(\mathbf{x}(t)). \tag{8.40}$$

Accordingly, we have the following theorem.

Theorem 8.5 (See Yu et al. [141]): *When adopting the activation function (8.39), the equilibrium of the resulting ZND (PTS-ZND) model (8.22), where the convergence parameter is set as*

$$\xi = \frac{\arctan\left(\mathbf{x}^{\mathrm{T}}(0)\mathbf{x}(0)\right)^{\frac{\rho}{2}}}{\rho t_{\mathrm{p}}} \tag{8.41}$$

with t_{p} being predefined convergence time, is PTS stable.

Proof: Based on (8.39), $\forall \mathbf{x}(t) \neq 0$, we have

$$\begin{aligned}\frac{\dot{\mathbf{x}}(t)}{\|\dot{\mathbf{x}}(t)\|_2} &= -\frac{\|\mathbf{x}(t)\|_2^{\rho}\mathbf{x}(t) + \|\mathbf{x}(t)\|_2^{-\rho}\mathbf{x}(t)}{\left\|\|\mathbf{x}(t)\|_2^{\rho}\mathbf{x}(t) + \|\mathbf{x}(t)\|_2^{-\rho}\mathbf{x}(t)\right\|_2} \\ &= \frac{-\mathbf{x}(t)}{\|\mathbf{x}(t)\|_2}.\end{aligned} \tag{8.42}$$

According to Definition 8.1, we can conclude that the state $\mathbf{x}(t)$ of the PTS-ZND is ratio-persistent. Next, to analyze the stability and convergence of the PTS-ZND model, a Lyapunov candidate is defined as

$$l(t) = \mathbf{x}^{\mathrm{T}}(t)\mathbf{x}(t) \geq 0. \tag{8.43}$$

For $\mathbf{x}(t) \neq 0$, taking the first time derivative of the candidate leads to

$$\begin{aligned}\dot{l}(t) &= 2\mathbf{x}^{\mathrm{T}}(t)\dot{\mathbf{x}}(t) \\ &= -2\xi\mathbf{x}^{\mathrm{T}}(t)(\mathbf{sig}_{\mathrm{n}}^{1+\rho}(\mathbf{x}(t)) + \mathbf{sig}_{\mathrm{n}}^{1-\rho}(\mathbf{x}(t))) \\ &= -2\xi\left(\|\mathbf{x}(t)\|_2^{2+\rho} + \|\mathbf{x}(t)\|_2^{2-\rho}\right) \\ &= -2\xi\left(l^{1+\frac{\rho}{2}}(t) + l^{1-\frac{\rho}{2}}(t)\right) \\ &< 0.\end{aligned} \tag{8.44}$$

According to the Lyapunov's second method, the PTS-ZND model is stable. By defining $\upsilon(t) = l^{\frac{\rho}{2}}(t)$, we have

$$\begin{aligned}\dot{\upsilon}(t) &= \frac{\rho}{2}l^{\frac{\rho}{2}-1}(t)\dot{l}(t) \\ &= -\xi\rho l^{\frac{\rho}{2}-1}\left(l^{1+\frac{\rho}{2}}(t) + l^{1-\frac{\rho}{2}}(t)\right) \\ &= -\xi\rho\left(l^{\rho} + 1\right) \\ &= -\rho\xi\left(\upsilon^2(t) + 1\right).\end{aligned} \tag{8.45}$$

Therefore, the settling time is obtained as

$$\begin{aligned}t_{\mathrm{s}} &= \frac{1}{\rho\xi}\int_0^{\upsilon(0)}\frac{1}{1+\upsilon^2}\mathrm{d}\upsilon \\ &= \frac{1}{\rho\xi}\arctan(\upsilon(0)) \\ &= t_{\mathrm{p}}.\end{aligned} \tag{8.46}$$

These results imply that the Lyapunov candidate and system state $\mathbf{x}(t)$ converge to zero in predefined-time t_{p}. Therefore, the PTS-ZND model is predefined-time stable. Based on the above analysis and Definition 8.3, we can conclude that the PTS-ZND model is PTS stable.

Remark 8.1: There are two limitations of conventional ZND approaches. First, the convergence of the control method based on the L-ZND model cannot meet the requirements of some tasks, where each robot needs to complete a sub-task within a predefined time. Second, existing nonlinear ZND approaches cannot ensure the synchronization of all dimensions in the task space error. The PTS-ZND model features the nonlinear activation function. On the one hand, it ensures the ratio persistence property (that is, the synchronization of all dimensions of the system state) by using the norm-normalized sign function. On the other hand, its special structure ensures that the system state converges to zero within a predefined time. Therefore, the PTS-ZND model is able to address the limitations of conventional ZND approaches.

8.3.3 Model-free Control

This section introduces a model-free control method for continuum robots based on the above PTS-ZND model.

8.3.3.1 PTS-ZND for Inverse Kinematics

First, the PTS-ZND model is adopted to solve the inverse kinematics problem. A vector-valued error function is firstly defined as

$$\mathbf{e} = \mathbf{p}_{\mathrm{d}} - \mathbf{p}. \tag{8.47}$$

According to the design formula of the ZND model (8.22), we have

$$\dot{\mathbf{e}} = -\xi \mathbf{g}(\mathbf{e}), \tag{8.48}$$

where $\mathbf{g}$ is the activation function (8.39) associated with the PTS-ZND model. Substituting $\mathbf{e}$ and (8.11) into (8.48) leads to

$$\dot{\mathbf{p}}_{\mathrm{d}} - \mathcal{J}\dot{\mathbf{u}} = -\xi \mathbf{g}(\mathbf{p}_{\mathrm{d}} - \mathbf{p}), \tag{8.49}$$

where $\dot{\mathbf{p}}_{\mathrm{d}}$ is the desired velocity. Then, the least-squares solution to (8.49) is obtained as follows:

$$\dot{\mathbf{u}} = \mathcal{J}^{\dagger} \left(\dot{\mathbf{p}}_{\mathrm{d}} + \xi \mathbf{g}(\mathbf{p}_{\mathrm{d}} - \mathbf{p})\right), \tag{8.50}$$

where $\mathcal{J}^{\dagger}$ denotes the pseudo-inverse of the Jacobian matrix. To ensure numerical stability, the pseudo-inverse can be calculated as

$$\mathcal{J}^{\dagger} = \mathcal{J}^{\mathrm{T}}(\mathcal{J}\mathcal{J}^{\mathrm{T}} + \lambda I)^{-1}, \tag{8.51}$$

where $\lambda \geq 0$ is a damping factor and I is an identity matrix. For a velocity-controlled continuum robot, the velocity $\dot{\mathbf{u}}$ calculated based on (8.50) is directly used to control the continuum robot. Considering that some continuum robots may be controlled by actuation magnitude, one can further calculate $\mathbf{u}$ by utilizing the Euler's forward formula:

$$\mathbf{u}_{k+1} \leftarrow \mathbf{u}_k + \tau \dot{\mathbf{u}}_k, \tag{8.52}$$

where τ denotes the sampling interval of time and the subscript k indicates the values of relevant variables at the kth time instant $t = k\tau$.

8.3.3.2 IGD-based Kinematics Learning

In this part, we seek to learn the unknown Jacobian $\mathcal{J}$ based on the input–output data of continuum robots. According to (8.11), we have

$$\Delta\mathbf{p} \approx \mathcal{J}\Delta\mathbf{u}, \tag{8.53}$$

where $\Delta\mathbf{u}$ denotes the incremental amount of actuation signal and $\Delta\mathbf{p}$ denotes the resulting end-effector displacement. Then, the following norm-based error function is defined:

$$e_{\mathrm{v}} = \frac{1}{2}\|\Delta\mathbf{p} - \hat{\mathcal{J}}\Delta\mathbf{u}\|_2^2, \tag{8.54}$$

where $\hat{\mathcal{J}}$ is the unknown Jacobian to be learned, and $\hat{\mathcal{J}}\Delta\mathbf{u}$ can be deemed as the estimated end-effector displacement. Next, to ensure the error function (8.54) evolves along the negative gradient direction, the learning law of the unknown Jacobian is designed as follows:

$$\Delta\hat{\mathcal{J}} = -\gamma\frac{\partial e_{\mathrm{v}}}{\partial \hat{\mathcal{J}}} = \gamma\left(\Delta\mathbf{p} - \hat{\mathcal{J}}\Delta\mathbf{u}\right)\Delta\mathbf{u}^{\mathrm{T}}, \tag{8.55}$$

where $\gamma > 0$ denotes the learning rate. Besides, an initial value of the unknown Jacobian needs to be determined for (8.55). This objective is achieved by changing the ith $(i = 1, 2, \cdots, n)$ element of the actuation signal by a sufficiently small incremental amount Δu_i independently and measuring its effect on the end-effector, i.e., the end-effector displacement $\Delta\mathbf{p}_i \in \mathbb{R}^m$. Then, at the time instant t_0, the value of the Jacobian matrix is initialized as

$$\hat{\mathcal{J}}_0 = \left[\frac{\Delta\mathbf{p}_1}{\Delta u_1}, \frac{\Delta\mathbf{p}_2}{\Delta u_2}, \cdots, \frac{\Delta\mathbf{p}_n}{\Delta u_n}\right]. \tag{8.56}$$

Finally, the unknown Jacobian matrix is updated as

$$\hat{\mathcal{J}}_{k+1} = \hat{\mathcal{J}}_k + \Delta\hat{\mathcal{J}}_k. \tag{8.57}$$

Generally, the learning law (8.55) and the Jacobian updating are performed only once during a time step, which may lead to insufficient accuracy to the learned Jacobian matrix. Therefore, we further present an IGD method to improve

Algorithm 3: Description of control algorithm.

Setting: Task duration T_{d}, time interval τ, reference path $\mathbf{p}_{\mathrm{d}}$, maximum iteration number N, initial joint angle $\mathbf{u}_0$.
Initialize $\hat{\mathcal{J}}_0$:
for $i = 1 : n$ **do**
  Control the continuum robot with $\mathbf{u}_0$
  Measure the end-effector position $\mathbf{p}_0$
  $\mathbf{u}_i = \mathbf{u}_0 + \mathbf{w}_i \Delta u_i$ (where $\mathbf{w}_i$ is a unit vector with the ith element being 1)
  Control the continuum robot with $\mathbf{u}_i$
  Measure the end-effector position $\mathbf{p}_i$
  Initialize the ith column of $\hat{\mathcal{J}}_0$ as $(\mathbf{p}_i - \mathbf{p}_0)/\Delta u^i$
end for
Start tracking:
for $k = 0 : T_{\mathrm{d}}/\tau$ **do**
  Get the actuation signal $\mathbf{u}_k$
  Measure the end-effector position $\mathbf{p}_k$
  if $k > 0$ **then**
    $i = 0$
    $\hat{J}_i = \hat{\mathcal{J}}_{k-1}$
    $\Delta \mathbf{p}_k = \mathbf{p}_k - \mathbf{p}_{k-1}$
    $\Delta \mathbf{u}_k = \mathbf{u}_k - \mathbf{u}_{k-1}$
    for $i = 0 : N - 1$ **do**
      $\Delta \hat{J}_i = \gamma \left(\Delta \mathbf{p}_k - \hat{J}_i \Delta \mathbf{u}_k \right) \Delta \mathbf{u}_k^{\mathrm{T}}$
      $\hat{J}_{i+1} = \hat{J}_i + \Delta \hat{J}_i$
    end for
    Update Jacobian $\hat{\mathcal{J}}_k = \hat{J}_N$
  end if
  PTS-ZND: $\dot{\mathbf{u}}_k = \hat{\mathcal{J}}_k^{\dagger} \left(\dot{\mathbf{p}}_{\mathrm{d},k} + \xi \mathbf{g}(\mathbf{p}_{\mathrm{d},k} - \mathbf{p}_k) \right)$
  Control the continuum robot with $\dot{\mathbf{u}}_k$
end for

accuracy. Specifically, the learning law and Jacobian updating are performed iteratively in the kth time step until it reaches the maximum iteration number N.

For clarity, the whole inverse kinematics method based on the PTS-ZND model and IGD method is elaborated in Algorithm 3 and illustrated in Figure 8.1.

Remark 8.2: In this remark, we analyze the connection between kinematics learning and predefined-time performance. To begin with, we define the Lyapunov candidate as

$$l(t) = \mathbf{e}^{\mathrm{T}} \mathbf{e}. \tag{8.58}$$

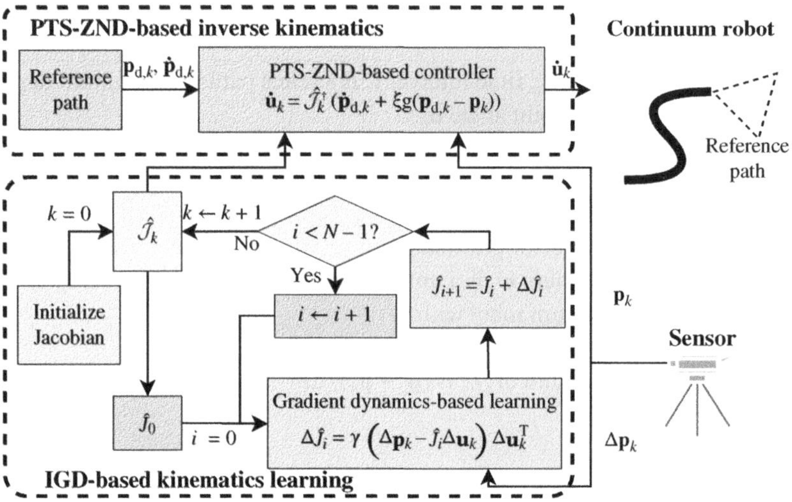

Figure 8.1 Block diagram of control method.

Taking the first time derivative of the candidate leads to

$$\dot{l}(t) = 2\mathbf{e}^{\mathrm{T}}(\dot{\mathbf{p}}_{\mathrm{d}} - \mathcal{J}\dot{\mathbf{u}}), \tag{8.59}$$

where

$$\dot{\mathbf{u}} = \hat{\mathcal{J}}^{\dagger}\,(\dot{\mathbf{p}}_{\mathrm{d}} + \xi\mathbf{g}(\mathbf{e}))\,. \tag{8.60}$$

By denoting the error between the real Jacobian matrix $\mathcal{J}$ and the estimated Jacobian matrix $\hat{\mathcal{J}}$ as $\Delta\mathcal{J} = \hat{\mathcal{J}} - \mathcal{J}$, we further get

$$\dot{l}(t) = -2\xi\mathbf{e}^{\mathrm{T}}\mathbf{g}(\mathbf{e}) + 2\mathbf{e}^{\mathrm{T}}\Delta\mathcal{J}\dot{\mathbf{u}}. \tag{8.61}$$

If the unknown Jacobian matrix can be accurately estimated (i.e., $\Delta\mathcal{J} = 0$), following the above proof, we can infer that the tracking error converges to zero before the predefined time t_{p}.

8.4 Prescribed-performance Control Method

This section elaborates the theoretical basis for the PPC of HCRs. First, the PPC theory is presented, followed by an inverse kinematics solver designed based on the PPC theory and ZND. Then, the design of Jacobian estimator is presented. Finally, the approach for convergence enhancement is introduced.

8.4.1 Prescribed Performance

One of the abovementioned challenges is to prespecify the task-space behavior of continuum robots. In particular, this chapter introduces how to achieve prescribed transient and steady performance on the error-monitoring function $e(t)$ for continuum robots. According to the literature [129], a smooth performance function $\rho : \mathbb{R}^+ \rightarrow \mathbb{R}^+$ should be firstly defined, which satisfies: 1) $\rho(t)$ is positive and decreasing; 2) $\lim_{t\rightarrow\infty} \rho(t) = \rho_\infty$. An example of the performance function is defined as

$$\rho(t) = (\rho_0 - \rho_\infty)\exp(-ct) + \rho_\infty, \tag{8.62}$$

where ρ_0, ρ_∞, and c are appropriately defined positive constants. To achieve prescribed performance is equivalent to

$$\begin{cases} -\delta\rho(t) < e(t) < \rho(t), & e(0) \geq 0 \\ -\rho(t) < e(t) < \delta\rho(t), & e(0) < 0 \end{cases} \tag{8.63}$$

where $0 \leq \delta \leq 1$. The graphical illustration of (8.63) is depicted in Figure 8.2.

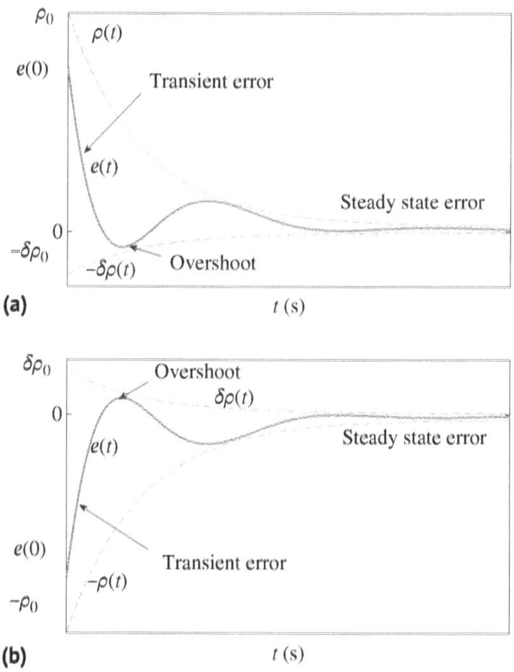

Figure 8.2 Illustration of prescribed tracking error: (a) $e(0) \geq 0$; (b) $e(0) < 0$.

To achieve the control of robots with prescribed performance, the error transformation step is necessary, which transforms the original system with constrained tracking error (8.63) into equivalent unconstrained one. Specifically, the error-monitoring function is defined as follows:

$$e(t) = \rho(t)h(\varepsilon(t)), \tag{8.64}$$

where $\varepsilon(t)$ is the transformed error and $h(\cdot)$ is a smooth and strictly increasing function defining the following bijective mapping:

$$\begin{cases} h : (-\infty, \infty) \rightarrow (-\delta, 1), & e(0) \geq 0 \\ h : (-\infty, \infty) \rightarrow (-1, \delta), & e(0) < 0 \end{cases}. \tag{8.65}$$

In particular, a candidate transformation function is

$$h(\varepsilon) = \begin{cases} \dfrac{\exp(\varepsilon) - \delta \exp(-\varepsilon)}{\exp(\varepsilon) + \exp(-\varepsilon)}, & e(0) \geq 0 \\ \dfrac{\delta \exp(\varepsilon) - \exp(-\varepsilon)}{\exp(\varepsilon) + \exp(-\varepsilon)}, & e(0) < 0 \end{cases}. \tag{8.66}$$

Considering the definitions of $\rho(t)$ and $h(\cdot)$, we have the following well-defined inverse transformation:

$$\varepsilon(t) = h^{-1}\left(\frac{e(t)}{\rho(t)}\right). \tag{8.67}$$

It has been shown that keeping $\varepsilon(t)$ bounded $\forall t \geq 0$ is equivalent to guaranteeing (8.63) [129], thereby achieving prescribed performance on the error-monitoring function $e(t)$.

8.4.2 Robot Kinematics

Typically, the forward kinematics of a continuum robot is given by

$$\mathbf{p}(t) = \mathbf{f}(\mathbf{u}(t)), \tag{8.68}$$

where $\mathbf{p}(t) \in \mathbb{R}^m$ denotes the position of robot end-effector in the m-dimensional task space, $\mathbf{u}(t) \in \mathbb{R}^n$ denotes the state of n actuators of the continuum robot, and $\mathbf{f}(\cdot)$ denotes a nonlinear kinematic mapping function. Accordingly, the differential kinematics of the continuum robot is as follows:

$$\dot{\mathbf{p}}(t) = J(t)\dot{\mathbf{u}}(t), \tag{8.69}$$

where $\dot{\mathbf{p}}(t)$ denotes the first time derivative of $\mathbf{p}(t)$ and $J(t) = \partial \mathbf{f}(\mathbf{u}(t))/\partial \mathbf{u}(t) \in \mathbb{R}^{m\times n}$ is the Jacobian matrix of the continuum robot.

8.4.3 Inverse Kinematics Solver

To begin with, the desired task-space state is denoted by $\mathbf{p}_{\mathrm{d}}(t) \in \mathbb{R}^m$, which is manually specified by users or determined by planning algorithms. For a continuum robot, a vector-valued error-monitoring function is defined as

$$\mathbf{e}(t) = \mathbf{p}_{\mathrm{d}}(t) - \mathbf{p}(t). \tag{8.70}$$

The objective of this chapter is to achieve predefined behavior (i.e., prescribed performance) on the error-monitoring function (8.70), which is formulated as

$$\begin{cases} -\delta\rho_i(t) < e_i(t) < \rho_i(t), & e_i(0) \geq 0 \\ -\rho_i(t) < e_i(t) < \delta\rho_i(t), & e_i(0) < 0 \end{cases}, \tag{8.71}$$

where $\boldsymbol{\rho}(t) = [\rho_1, \rho_2, \cdots, \rho_m]^{\mathrm{T}}$ denotes the vector of performance functions, δ is a constant parameter, and the subscript $i = 1, 2, \cdots, m$ indicates the ith element of related vectors. The objective can be achieved by utilizing the error transformation method and keeping the transformed error bounded. The constrained error-monitoring function $\mathbf{e} = [e_1, e_2, \cdots, e_m]^{\mathrm{T}}$ can be transformed into equivalent unconstrained one $\boldsymbol{\varepsilon} = [\varepsilon_1, \varepsilon_2, \cdots, \varepsilon_m]^{\mathrm{T}}$ by using the following transformation:

$$e_i(t) = \rho_i(t)h_i(\varepsilon_i(t)) \tag{8.72}$$

with $\mathbf{h}(\cdot) = [h_1, h_2, \cdots, h_m]^{\mathrm{T}}$ denotes the vector of m smooth and strictly increasing bijective mapping functions defined in (8.65) and (8.66). Accordingly, the following well-defined inverse transformation is obtained:

$$\varepsilon_i(t) = h_i^{-1}(z_i(t)), \tag{8.73}$$

where

$$z_i(t) = \frac{e_i(t)}{\rho_i(t)}. \tag{8.74}$$

Next, the objective is to keep each element of $\boldsymbol{\varepsilon}(t)$ bounded. For any bounded initial task-space error $e_i(0)$, it is always possible to find $\rho_i(0)$ satisfying $0 \leq |e_i(0)| < \rho_i(0)$, which means $\varepsilon_i(0)$ is bounded. Considering the property of ZND, it can be utilized to drive $\varepsilon_i(t)$ to zero, thereby keeping $\varepsilon_i(t)$ bounded. Therefore, the following vector-valued ZND model is designed:

$$\dot{\boldsymbol{\varepsilon}}(t) = -\lambda\boldsymbol{\phi}(\boldsymbol{\varepsilon}(t)), \tag{8.75}$$

where $\lambda \in \mathbb{R}^+$ is a parameter and $\boldsymbol{\phi} = [\phi_1, \phi_2, \cdots, \phi_m]^{\mathrm{T}}$ is an array of monotonically increasing activation functions. By expanding the above ZND model with (8.73) and (8.74), the dynamic equation of the ith subsystem is attained as follows:

$$\frac{\mathrm{d}h_i^{-1}}{\mathrm{d}z_i} \cdot \frac{\mathrm{d}z_i}{\mathrm{d}t} = -\lambda\phi_i\left(h_i^{-1}(z_i)\right), \tag{8.76}$$

where

$$h_i^{-1}(z_i) = \begin{cases} \frac{1}{2}\ln\frac{\delta + z_i}{1 - z_i}, & e_i(0) \geq 0 \\ \frac{1}{2}\ln\frac{1 + z_i}{\delta - z_i}, & e_i(0) < 0 \end{cases}, \tag{8.77}$$

$$\frac{\mathrm{d}h_i^{-1}}{\mathrm{d}z_i} = \begin{cases} \frac{\delta + 1}{2(\delta + z_i)(1 - z_i)}, & e_i(0) \geq 0 \\ \frac{\delta + 1}{2(1 + z_i)(\delta - z_i)}, & e_i(0) < 0 \end{cases}, \tag{8.78}$$

$$\frac{\mathrm{d}z_i}{\mathrm{d}t} = \frac{(\dot{p}_{\mathrm{d},i} - \dot{p}_i)\rho_i - (p_{\mathrm{d},i} - p_i)\dot{\rho}_i}{\rho_i^2}. \tag{8.79}$$

Then, the compact form of all the subsystems is written as

$$\dot{\mathbf{p}}_{\mathrm{d}} - J\dot{\mathbf{u}} - \Lambda_\rho(\mathbf{p}_{\mathrm{d}} - \mathbf{p}) = -\lambda\Lambda_h^{-1}\boldsymbol{\phi}\left(\mathbf{h}^{-1}(\mathbf{z})\right), \tag{8.80}$$

where Λ_ρ and Λ_h are diagonal matrices defined as follows:

$$\Lambda_\rho = \mathrm{Diag}\left\{\frac{\dot{\rho}_1}{\rho_1}, \frac{\dot{\rho}_2}{\rho_2}, \cdots, \frac{\dot{\rho}_m}{\rho_m}\right\}, \tag{8.81}$$

$$\Lambda_h = \mathrm{Diag}\left\{\frac{\mathrm{d}h_1^{-1}}{\rho_1\mathrm{d}z_1}, \frac{\mathrm{d}h_2^{-1}}{\rho_2\mathrm{d}z_2}, \cdots, \frac{\mathrm{d}h_m^{-1}}{\rho_m\mathrm{d}z_m}\right\}. \tag{8.82}$$

By solving the dynamic equation (8.80), one can obtain the required vector of actuation signal $\mathbf{u}$. However, the actuation space of continuum robots is usually bounded, whereas the limits have not been considered in the previous steps. Therefore, it is necessary to further impose constraints to the actuation signal. The constrained actuation signal vector $\mathbf{u}$ can be transformed into an unconstrained vector θ by utilizing the following element-wise transformation:

$$u_j = g_j(\theta_j) = \frac{\exp(\theta_j)(u_{\max,j} - u_{\min,j})}{\exp(\theta_j) + \exp(-\theta_j)} + u_{\min,j} \tag{8.83}$$

where $\mathbf{g} = [g_1, g_2, \cdots, g_n]^{\mathrm{T}}$ is the vector of n transformation functions, $j = 1, 2, \cdots, n$ indicates the jth element of related vectors, and $u_{\max,j}/u_{\min,j}$ denotes the upper/lower limit of actuation signal. Replacing $\mathbf{u}$ in (8.80) with $\mathbf{g}(\theta)$, the least-square solution is obtained as follows:

$$\dot{\theta} = M^{\mathrm{T}}\left(MM^{\mathrm{T}} + \gamma_{\mathrm{p}}^2 I_m\right)^{-1}\left(\dot{\mathbf{p}}_{\mathrm{d}} - \Lambda_\rho(\mathbf{p}_{\mathrm{d}} - \mathbf{p}) + \lambda\Lambda_h^{-1}\boldsymbol{\phi}\left(\mathbf{h}^{-1}(\mathbf{z})\right)\right) \tag{8.84}$$

where $\gamma_{\mathrm{p}} \in \mathbb{R}$ is a damping constant to ensure that the solution is numerical stable and well-behaved near singularities, $I_m \in \mathbb{R}^{m\times m}$ is an identity matrix, and $M = J\Lambda_g$ with

$$\Lambda_g = \mathrm{Diag}\left\{\frac{\mathrm{d}g_1}{\mathrm{d}\theta_1}, \frac{\mathrm{d}g_2}{\mathrm{d}\theta_2}, \cdots, \frac{\mathrm{d}g_n}{\mathrm{d}\theta_n}\right\}, \tag{8.85}$$

$$\frac{\mathrm{d}g_j}{\mathrm{d}\theta_j} = \frac{2(u_{\max,j} - u_{\min,j})}{\left(\exp(\theta_j) + \exp(-\theta_j)\right)^2}. \tag{8.86}$$

The unconstrained vector θ can be obtained by solving (8.84), and the constrained actuation signal vector u is calculated via (8.83). It is worth pointing out that only when the robot has more degrees of freedom than necessary to fulfill the end-effector task can it simultaneously fulfill the end-effector task and avoid actuation-space limits. Otherwise, the end-effector task cannot be achieved successfully.

8.4.4 Jacobian Estimator

It is evident that the solution of (8.84) depends on the Jacobian matrix J of continuum robots, which is difficult to be analytically and accurately calculated. Therefore, a Jacobian estimator is developed in this part to estimate the Jacobian matrix without knowing the analytical model and kinematic parameters of continuum robots.

First, the Jacobian matrix of a continuum robot is initialized through the following motor babbling procedures:

1) Changing the jth actuation signal of the robot by a sufficiently small increment amount $\Delta u_j \in \mathbb{R}^+$ independently;
2) Measuring and recording the state change $\Delta\mathbf{p}_j \in \mathbb{R}^m$ in the task space;
3) Determining the jth column of the Jacobian matrix $J_j = \dfrac{\Delta\mathbf{p}_j}{\Delta u_j} \in \mathbb{R}^m$.

By repeating the above procedures for the n actuation dimensions of the robot, the initial Jacobian matrix is finally determined as

$$J(0) = \left[\frac{\Delta\mathbf{p}_1}{\Delta u_1}, \frac{\Delta\mathbf{p}_2}{\Delta u_2}, \cdots, \frac{\Delta\mathbf{p}_n}{\Delta u_n}\right]. \tag{8.87}$$

Since the value of the Jacobian matrix is concerned with robot configuration, it is necessary to develop a Jacobian adaptation method to ensure the accuracy of the Jacobian matrix. Then, the following vector-valued error-monitoring function is defined:

$$\mathbf{e}_{\mathrm{v}}(t) = \dot{\mathbf{p}}(t) - J(t)\dot{\mathbf{u}}(t), \tag{8.88}$$

where $\dot{\mathbf{p}}(t)$ is the actual task-space velocity of the robot, $J(t)$ is the Jacobian matrix to be adapted, and $J(t)\dot{\mathbf{u}}(t)$ can be regarded as the estimated task-space velocity. To drive the error-monitoring function to zero, the ZND model is adopted again as follows:

$$\dot{\mathbf{e}}_{\mathrm{v}}(t) = -\lambda\boldsymbol{\phi}(\mathbf{e}_{\mathrm{v}}(t)). \tag{8.89}$$

By expanding the above model with (8.88), the following vector-valued dynamic equation is attained:

$$\ddot{\mathbf{p}} - \dot{J}\dot{\mathbf{u}} - J\ddot{\mathbf{u}} = -\lambda\boldsymbol{\phi}\,(\dot{\mathbf{p}} - J\dot{\mathbf{u}}), \tag{8.90}$$

where $\ddot{\mathbf{p}}$ denotes the task-space acceleration, $\dot{J}$ is the first time derivative of J, and $\ddot{\mathbf{u}}$ denotes the actuation-space acceleration. In (8.90), the unknown variable to be solved is $\dot{J}$. Considering the solution to the under-determined equation is not unique, the least-square solution is selected as the optimal solution. Therefore, the dynamic equation (8.90) is rewritten as

$$A\mathrm{vec}(\dot{J}^{\mathrm{T}}) = \mathbf{b}, \tag{8.91}$$

where $A = I_m \otimes \dot{\mathbf{u}}^{\mathrm{T}}$ denotes the Kronecker product of the identity matrix I_m and $\dot{\mathbf{u}}^{\mathrm{T}}$ [142], $\mathrm{vec}(\cdot)$ denotes the vectorization of a matrix, and

$$\mathbf{b} = \ddot{\mathbf{p}} - J\ddot{\mathbf{u}} + \lambda\boldsymbol{\phi}\,(\dot{\mathbf{p}} - J\dot{\mathbf{u}}). \tag{8.92}$$

Finally, the Jacobian matrix is adapted based on the following dynamic equation:

$$\mathrm{vec}(\dot{J}^{\mathrm{T}}) = A^{\mathrm{T}}(AA^{\mathrm{T}} + \gamma_{\mathrm{v}}^2 I_m)^{-1}\mathbf{b} \tag{8.93}$$

where $\gamma_{\mathrm{v}} \in \mathbb{R}$ is a damping constant.

8.4.5 Convergence Enhancement

By combining the behavior-prescribed controller and Jacobian estimator, the behavior-prescribed adaptive control (BPAC) system for continuum robots is obtained as illustrated in Figure 8.3 and elaborated in Algorithm 4. Several advantages of the control system are worth highlighting. First, the motor babbling process and the ZND model are respectively utilized to initialize and adapt the unknown Jacobian matrix of continuum robots, which is further used to support the operation of the controller. Therefore, the method avoids the complex modeling process and uncertainty of continuum robots. Second, users can specify the task-space behavior and performance of continuum robots in advance with the desired task-space state $\mathbf{p}_d(t)$ and performance function $\rho(t)$ by means of the task-space error transformation and the ZND model. Third, by utilizing the actuation-space transformation, it is ensured that the actuation signal generated by the controller satisfies the actuation space constraint, thereby protecting the actuators of continuum robots. Finally, it is not difficult to find that the design of the control system is independent of the types of continuum robots and tasks, which implies the control system is applicable to HCRs and different tasks (including end-effector kinematic control and visual servoing). Despite the abovementioned advantages, the control system can be further improved in terms of convergence and robustness by adopting appropriate performance functions and nonlinear activation functions for ZND models.

For example, with the exponential performance function (8.62) (EPF), the task-space error is exponentially convergent to the upper boundary ρ_∞. However, the convergence process theoretically takes infinitely long time, which is unacceptable in practical applications. As illustrated in Figure 8.4, this chapter's objective is to

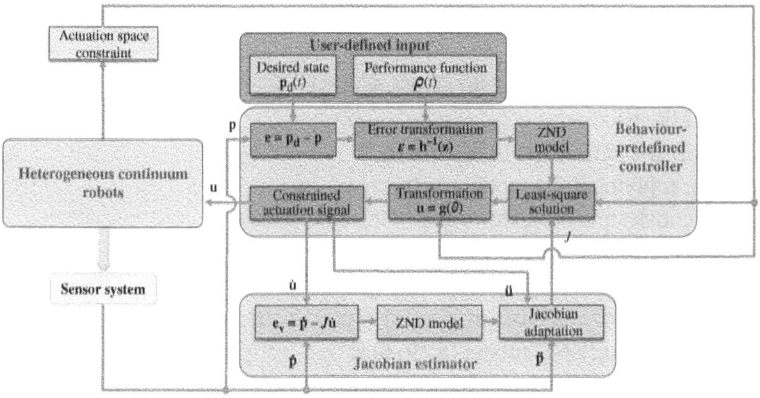

Figure 8.3 Diagram of behavior-prescribed adaptive control (BPAC) system for HCRs.

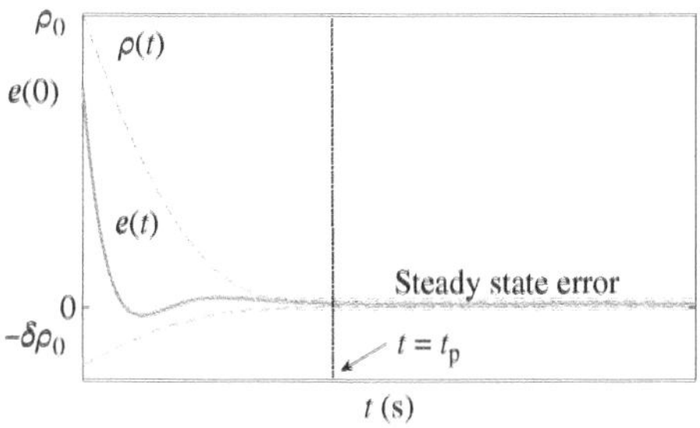

Figure 8.4 Illustration of prescribed performance control with predefined-time convergence, where task-space error $e(t)$ not only complies with performance function $\rho(t)$ but also converges to steady state before predefined time t_p.

Algorithm 4: General description regarding BPAC method for HCRs.

Input: Task duration T_{d}, initial actuation signal $\mathbf{u}(t_0)$, control interval τ, desired state $\mathbf{p}_{\mathrm{d}}(t)$, performance function $\rho(t)$, activation function $\boldsymbol{\phi}$, parameters.
Initialize Jacobian: $J(t_0)$
for $j = 1 : n$ **do**
    Control the robot with $\mathbf{u}_1 \leftarrow \mathbf{u}(t_0)$
    Measure the task-space state $\mathbf{p}_1$
    Modify the jth element of the actuation input $u_{1,j} \leftarrow u_{1,j} + \Delta u_j \in \mathbb{R}$
    Control the robot with $\mathbf{u}_1$
    Measure the task-space state $\mathbf{p}_2$
    Calculate the state change $\Delta \mathbf{p}_j \leftarrow \mathbf{p}_2 - \mathbf{p}_1 \in \mathbb{R}^m$
    Initialize the jth column of the Jacobian matrix $J_j(t_0) = \dfrac{\Delta \mathbf{p}_j}{\Delta u_j} \in \mathbb{R}^m$
end for
Execute task:
for $k = 0 : T_{\mathrm{d}}/\tau$ **do**
    Control the robot with $\mathbf{u}(t_k)$
    Obtain task-space feedback $\mathbf{p}(t_k)$, $\dot{\mathbf{p}}(t_k)$, $\ddot{\mathbf{p}}(t_k)$
    Define task-space error $\mathbf{e}(t_k) = \mathbf{p}_{\mathrm{d}}(t_k) - \mathbf{p}(t_k)$
    Error transformation $\boldsymbol{\varepsilon}(t_k) \leftarrow (8.72)$
    (8.80) $\leftarrow$ applying ZND model to $\boldsymbol{\varepsilon}(t_k)$
    Impose constraints by substituting (8.83) into (8.80)
    Obtain least-square solution $\dot{\boldsymbol{\theta}}(t_k)$ from (8.84)
    $\dot{\mathbf{u}}(t_k) = \Lambda_g \dot{\boldsymbol{\theta}}(t_k)$
    Define task-space velocity error $\mathbf{e}_{\mathrm{v}}(t_k) = \dot{\mathbf{p}}(t_k) - J(t_k)\dot{\mathbf{u}}(t_k)$
    (8.90) $\leftarrow$ applying ZND model (8.89) to $\mathbf{e}_{\mathrm{v}}(t_k)$
    Obtain least-square solution $\dot{J}(t_k)$ from (8.93)
    Unconstrained control signal $\boldsymbol{\theta}(t_{k+1}) = \boldsymbol{\theta}(t_k) + \tau\dot{\boldsymbol{\theta}}(t_k)$
    Update constrained control signal $\mathbf{u}(t_{k+1}) = \mathbf{g}(\boldsymbol{\theta}(t_{k+1}))$
    Update Jacobian $J(t_{k+1}) = J(t_k) + \tau\dot{J}(t_k)$
end for
Output: robot trajectory $\mathbf{p}(t)$

ensure that the task-space error not only complies with the performance function but also converges to steady state before the predefined time t_{p}. To achieve the PPC with predefined-time convergence, which means the task-space error should converge to an arbitrarily small upper boundary in predefined time, the following predefined-time performance function (PPF) can be adopted [129]:

$$\rho_i^{\mathrm{PPF}}(t) = \begin{cases} (\rho_{0,i} - \rho_{\infty,i}) \exp(\psi(t)) + \rho_{\infty,i}, & 0 \le t < t_{\mathrm{p}} \\ \rho_{\infty,i}, & t \ge t_{\mathrm{p}} \end{cases} \tag{8.94}$$

with

$$\psi(t) = -\frac{ct_{\mathrm{p}}t}{t_{\mathrm{p}} - t}, \tag{8.95}$$

where t_{p} is the convergence time specified by users.

8.5 Verification of PTC-ZND

In this section, simulations and experiments based on HCRs are performed to validate the predefined-time convergent control method.

8.5.1 Simulations

To begin with, simulations on a CCR and a CTR are performed to verify the effectiveness of the control scheme.

8.5.1.1 Settings

The control system is validated through simulations using a CCR and a CTR. The control method is employed to guide the CCR in tracking a circular path, and similarly, the CTR is also controlled to track a circular path. The durations of the tracking tasks are set to 10 s and 5 s, respectively. Furthermore, the parameter r of the PTC-ZND model remains constant at 0.9 throughout this chapter.

8.5.1.2 Cable-driven Continuum Robot

Some configurations of the CCR for tracking the circular path are illustrated in Figure 8.5. The CCR is composed of two connected segments. The actual trajectory of the robot end-effector is almost the same as the reference path. The profiles of tracking error and control signal of the CCR are depicted in Figure 8.6. It is found that the tracking error of the continuum robot almost converges to zero before $t = 3$ s, which satisfies the requirement ($t_{\mathrm{c}} = 3$ s). The steady-state tracking error of the CCR is less than 1×10^{-3} m during the tracking process. Besides, Figure 8.7 illustrates the comparison between the actual Jacobian matrix $J(t)$ and

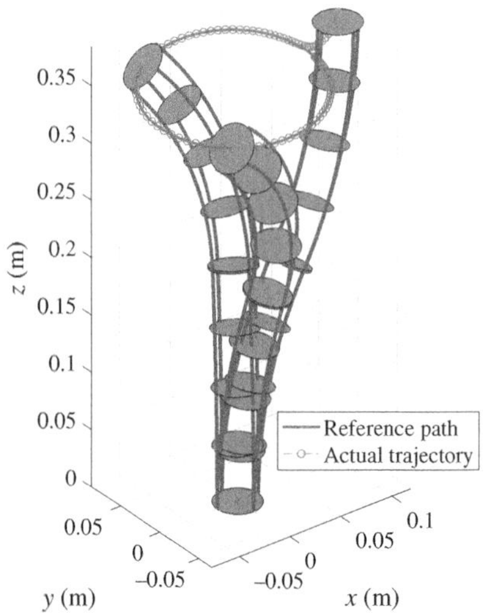

Figure 8.5 Some configurations of CCR tracking reference path in simulations.

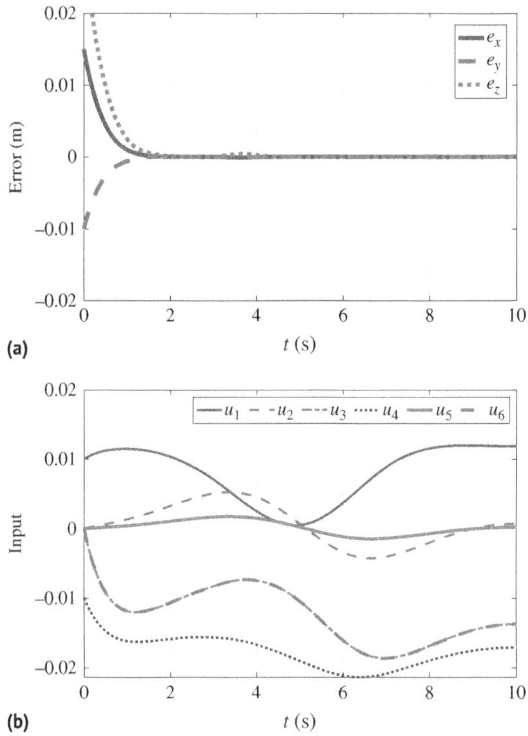

Figure 8.6 Simulation results of CCR for tracking reference path: (a) tracking errors; (b) control signals.

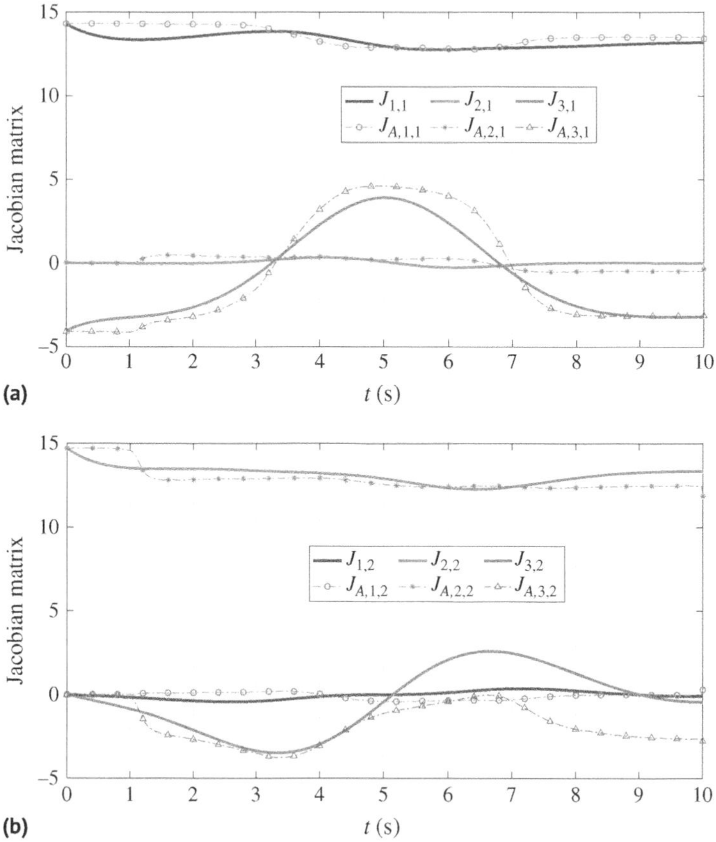

Figure 8.7 Comparison between actual Jacobian $J(t)$ and estimated Jacobian $J_A(t)$: (a) first column; (b) second column.

the estimated Jacobian matrix $J_A(t)$. It can be found that the estimated Jacobian elements can track the actual Jacobian matrix approximately, which reveals the effectiveness of the Jacobian estimator.

8.5.1.3 Concentric Tube Continuum Robot

Some configurations of the CTR when tracking the circular path are illustrated in Figure 8.8. The CTR consists of inner tube, mid tube, and outer tube. The actual trajectory of the robot end-effector is almost the same as the reference path. The profiles of related tracking error and control signal of the CTR are illustrated in Figure 8.9. The tracking error of the CTR almost converges to zero before $t = 1$ s, which also satisfies the requirement ($t_c = 1$ s).

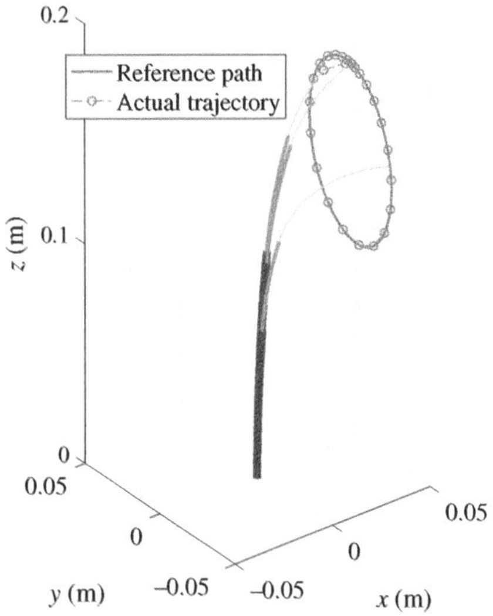

Figure 8.8 Some configurations of CTR tracking reference path in simulations.

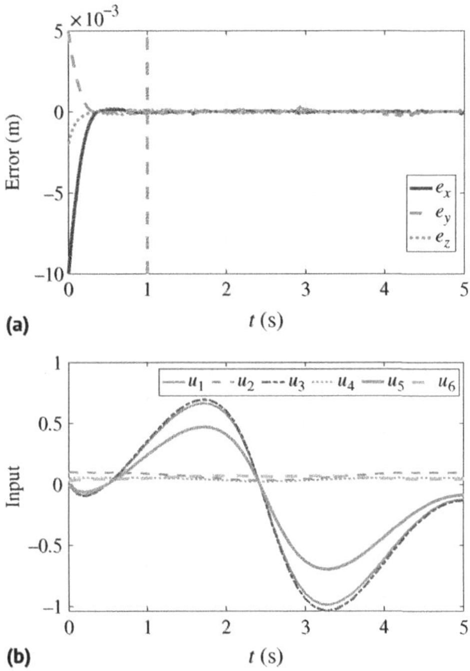

Figure 8.9 Simulation results of CTR for tracking reference path: (a) tracking errors; (b) control signals.

8.5.2 Experiments

In this section, physical experiments are carried out to verify the effectiveness of the control method.

8.5.2.1 Settings

The experimental setup consists of a CCR, a computer, and a Micron Tracker camera. The camera is utilized to accurately measure the position of the marker attached to the robot's end-effector. To estimate the end-effector velocity and acceleration, low-pass filters are employed. The robot is driven by motors, which are controlled directly by proportional-integral-differential (PID) controllers. The control system is implemented using Simulink, which sends actuation signals to the PID controllers, thereby actuating the motors. In this experiment, the CCR is controlled using the method to track a Lissajous path. The duration of the task is set to 60 s, with a predefined convergence time of $t_c = 3$ s.

8.5.2.2 Experiment Results

Experiment snapshots of the CCR tracking the Lissajous path are shown in Figure 8.10. The specific experiment results of the tracking task are illustrated in Figure 8.11. The figures show that the tracking error of the CCR almost converges to zero before $t = 3$ s, which satisfies the requirement ($t_c = 3$ s). Then, the end-effector moves along the reference path with the steady-state error less than 2×10^{-3} m.

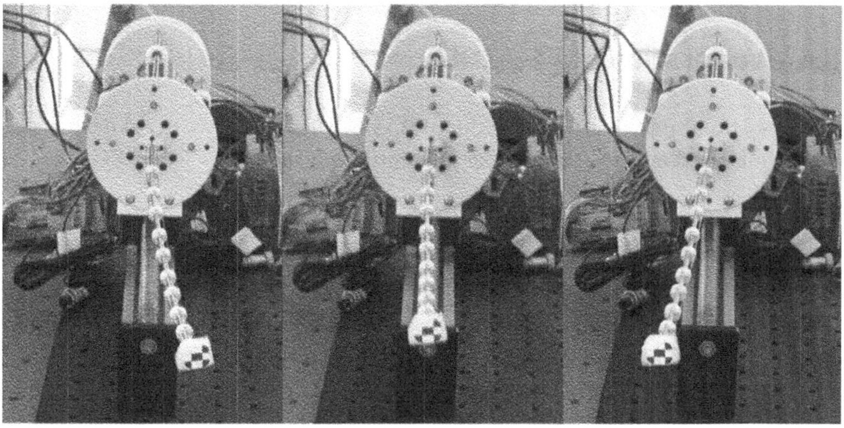

Figure 8.10 Experiment snapshots of CCR in tracking task.

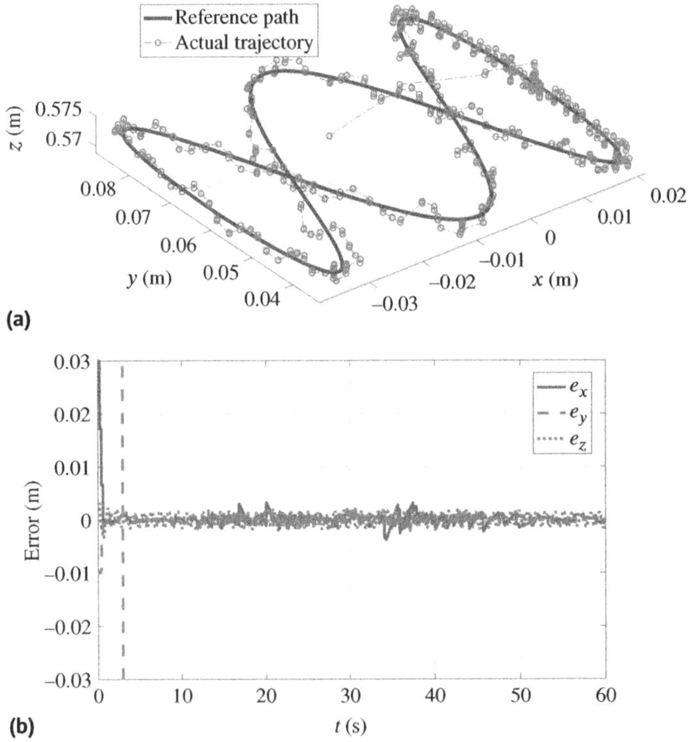

Figure 8.11 Experiment results of CCR for tracking reference path: (a) end-effector trajectories; (b) tracking errors.

8.6 Verification of PTS-ZND

In this section, simulations are conducted to verify the effectiveness of the control algorithm based on PTS-ZND model for continuum robots.

8.6.1 Setting

For simulations, a two-segment continuum robot is constructed. The maximum iteration number of the IGD was set to $N = 10000$. The parameter of the PTS-ZND model was set to $\rho = 0.6$. Our task was to control the continuum robot to track a circular path. The sampling interval of time was set to $\tau = 0.01$ s. In our simulations, to demonstrate the PTS convergence, the initial position error between the reference trajectory and the end-effector position was set as $\mathbf{e}_0 = [-0.02, -0.01, 0.01]^{\mathrm{T}}$ m.

8.6.2 Trajectory Tracking

As depicted in Figure 8.12, the continuum robot fulfilled the tracking task successfully, which reveals the feasibility of the control method. Figure 8.13 depicts more details about tracking results. Specially, the end-effector trajectory of the robot is expected to converge to the reference trajectory in $t_p = 5$ s. As illustrated in Figure 8.13b, the tracking errors with respect to different axes converge to zero before $t = 5$ s. At the same time, the tracking errors exhibit the property of time-synchronized convergence.

8.7 Verification of BPAC

In this section, simulations and physical experiments are performed to verify the effectiveness of the BPAC method.

8.7.1 Simulations

To begin with, simulative studies based on a CCR and a CTR are carried out to validate the BPAC method.

8.7.1.1 Behavior-predefined Control

To begin with, a simulation based on a two-segment fixed-base CCR is conducted to verify the feasibility of the BPAC method for prescribing the task-space behavior of continuum robots. The CCR is required to finish a star path tracking task with the task-space error constrained by the EPF. The task-space dimension is $n = 3$, and the actuation-space dimension is $m = 6$ for the CCR. The task duration is set to $T_d = 30$ s, and the control frequency is 100 Hz. The convergence parameter of the ZND model is set as $\lambda = 0.2$. A larger value of the convergence parameter leads to a

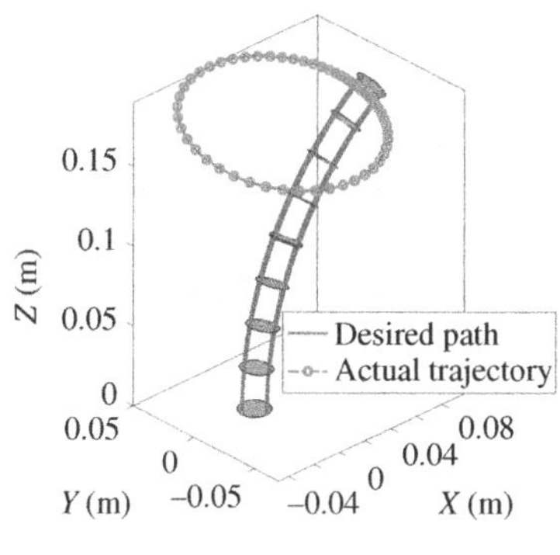

Figure 8.12 Illustration of continuum robot tracking circular path.

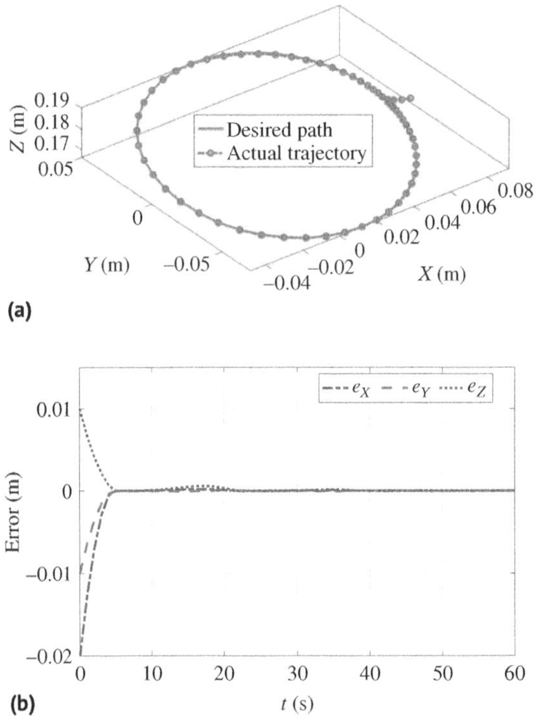

Figure 8.13 Tracking control results of continuum robot: (a) trajectory comparison; (b) position errors.

faster convergence rate of the ZND model, which, however, requires higher control frequency and more computing power. The initial position error between the robot end-effector and the reference path is $\mathbf{e}(0) = [20, 15, -15]^{\mathrm{T}}$ mm. The parameters of the EPF are set as $c = 0.2$, $\delta = 0.2$, $\boldsymbol{\rho}_0 = [40, 30, 30]^{\mathrm{T}}$ mm, and $\boldsymbol{\rho}_\infty = [1, 1, 1]^{\mathrm{T}}$ mm. c is used to adjust the convergence rate of EPF and tracking error $\mathbf{e}(t)$. δ is used to adjust the overshoot of tracking error. $\boldsymbol{\rho}_0$ is selected to satisfy $\rho_{0,i} > |e_i(0)|$. $\boldsymbol{\rho}_\infty$ specifies the value of EPF and the boundaries of tracking error when $t \to \infty$.

Some snapshots of the CCR performing the tracking task are displayed in Figure 8.14, which illustrates that the actual trajectory of robot end-effector converges to the reference path gradually. The profiles of the EPF and the task-space error with regard to different dimensions are illustrated in Figure 8.15a, from which one can find that the EPF $\boldsymbol{\rho}(t)$ is convergent to $\boldsymbol{\rho}_\infty$ exponentially and the task-space error $\mathbf{e}$ stays within the boundaries specified by the EPF. At the end of the task, we have $\boldsymbol{\rho}(T_{\mathrm{d}}) = [1.097, 1.072, 1.072]^{\mathrm{T}}$ mm and $\mathbf{e}(T_{\mathrm{d}}) = [0.4013, 0, 4168, -0.4072]^{\mathrm{T}}$ mm. These results have shown the feasibility of the

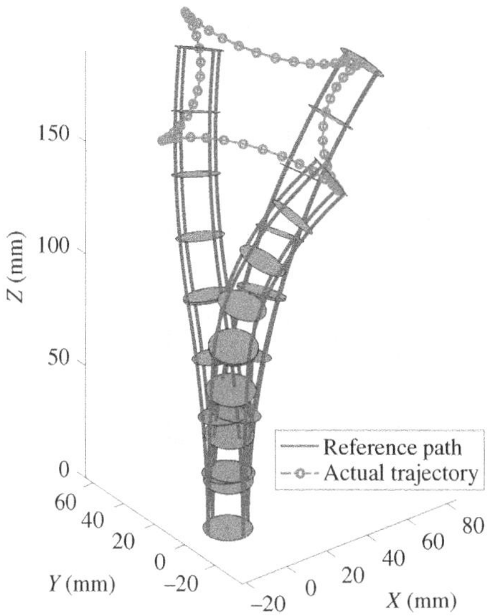

Figure 8.14 Some configurations of CCR performing star path tracking task in simulative studies.

BPAC method in prescribing the task-space behavior of continuum robots, including transient error and steady-state error. Besides, the numerical results have also revealed the disadvantage of EPF in terms of convergence. It can be seen that $\boldsymbol{\rho}(T_{\mathrm{d}})$ is not exactly equal to $\boldsymbol{\rho}_{\infty}$ because it theoretically takes infinitely long time for EPF to converge to $\boldsymbol{\rho}_{\infty}$, whereas the task duration is usually finite. Therefore, one can utilize or design better performance functions to improve the task-space convergence, which is studied below.

8.7.1.2 Comparison of Performance Functions

In this part, the influence of two different performance functions on the control system is analyzed. Although the EPF adopted in the above studies is exponentially convergent, it theoretically takes infinitely long time for EPF to converge to $\boldsymbol{\rho}_{\infty}$, while the task duration is usually finite in practical applications. Therefore, it is difficult to directly prescribe the time required for the performance function to converge to an arbitrarily small value when adopting the EPF. Considering the above shortcoming, this chapter adopts the PPF with $c = 0.2$ to perform the same tracking task and records the result. As illustrated in Figure 8.15b, the task-space error converges to $\boldsymbol{\rho}_{\infty} = [1, 1, 1]^{\mathrm{T}}$ mm before $t_{\mathrm{p}} = 10$ s. Therefore, compared with adopting EPF, using PPF provides the opportunity to directly prescribe the

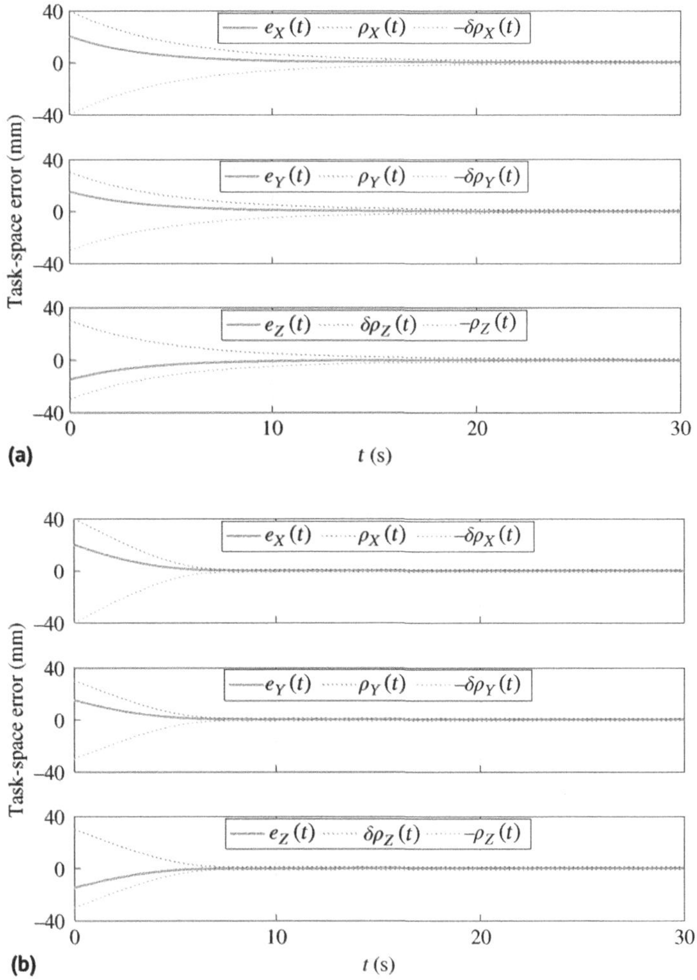

Figure 8.15 Profiles of $\rho(t) = [\rho_X, \rho_Y, \rho_Z]^T$ and task-space error $\mathbf{e}(t) = [e_X, e_Y, e_Z]^T$ of CCR when performing star path tracking task using (a) EPF and (b) PPF in simulative studies.

time required for the performance function and task-space error to converge to an arbitrarily small value.

8.7.1.3 CTR Control

In this part, simulations based on a CTR are performed to verify the adaptability of the BPAC method. The CTR is composed of an inner tube, a mid tube, and an outer tube, each of which can be rotated and translated. The CTR is controlled to track a circular path with the task-space error constrained by PPF. Some snapshots of the CTR tracking the reference path are depicted in Figure 8.16, which shows that the end-effector path converges to the reference path. The profiles of the PPF and the task-space error with regard to different dimensions are depicted in Figure 8.17, from which the PPF $\boldsymbol{\rho}(t)$ converges to $\boldsymbol{\rho}_\infty$ in predefined time ($t_p = 10$ s) and the task-space error $\mathbf{e}$ stays within the boundaries specified by the PPF. These results have revealed the feasibility of the BPAC method in controlling the CTR and prescribing its task-space performance.

8.7.2 Experiments

In this section, physical experiments are performed on five HCRs to validate the BPAC algorithm.

8.7.2.1 Cable-driven Continuum Robot

In this part, physical experiments based on a CCR are carried out. The CCR is required to track a circular path with its end-effector in the working space. The task duration is set as $T_d = 120$ s, and the control frequency is 10 Hz. The PPF is adopted for this task, and the convergence time is set as $t_p = 10$ s. The initial position error between the robot end-effector and the reference path is $\mathbf{e}(0) = [30, -10, -20]^T$

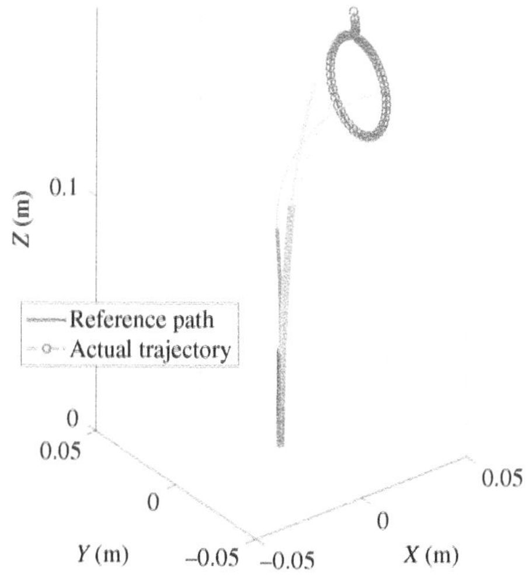

Figure 8.16 Some configurations of CTR performing circular path tracking task in simulative studies.

mm. The parameters of the PPF are set as $c = 0.2$, $\delta = 1$, $\boldsymbol{\rho}_0 = [45, 15, 30]^\mathrm{T}$ mm, and $\boldsymbol{\rho}_\infty = [2, 2, 2]^\mathrm{T}$ mm, which means the task-space error $\mathbf{e}(t)$ should converge to $\boldsymbol{\rho}_\infty$ in 10 s. The convergence parameter of the ZND models is set as $\lambda = 1$. Some snapshots of the CCR tracking the circular path are displayed in Figure 8.18. The comparison between the reference path and the actual trajectory is illustrated in Figure 8.19, and the task-space error is shown in Figure 8.20. It can be seen that the performance function and the task-space error converge to $\boldsymbol{\rho}_\infty$ before t_p as expected, and the end-effector moves along the reference path with the steady-state error less than 2 mm. These experiment results have verified the feasibility

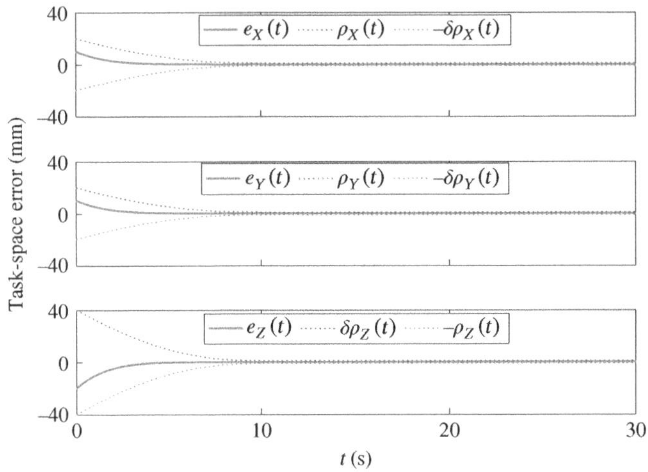

Figure 8.17 Profiles of PPF $\rho(t)$ and task-space error $\mathbf{e}(t)$ along different axes when CTR performs circular path tracking task in simulative studies.

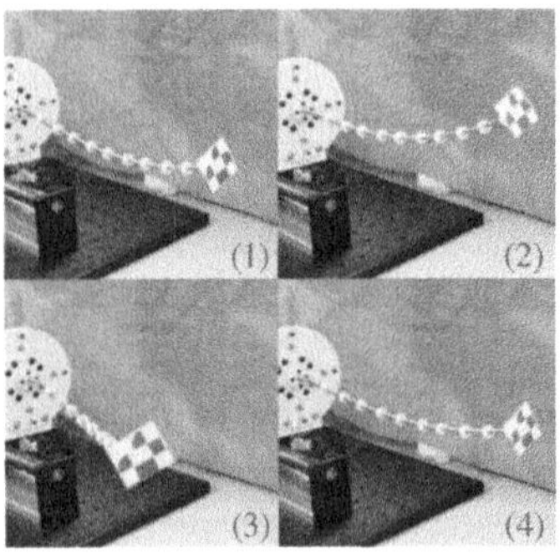

Figure 8.18 Snapshots of cable-driven continuum robot.

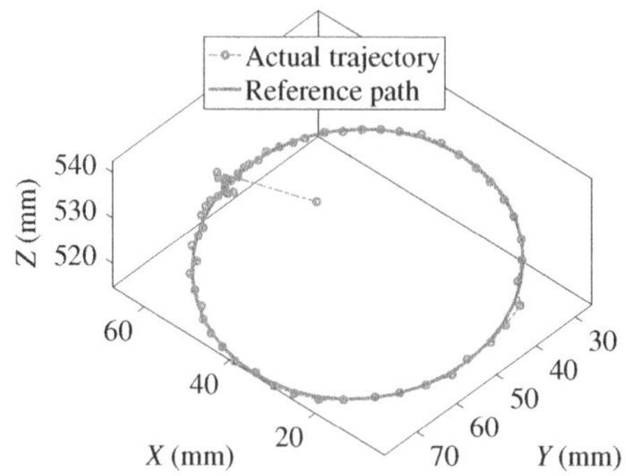

Figure 8.19 Trajectories of CCR performing circular path tracking task in experimental studies.

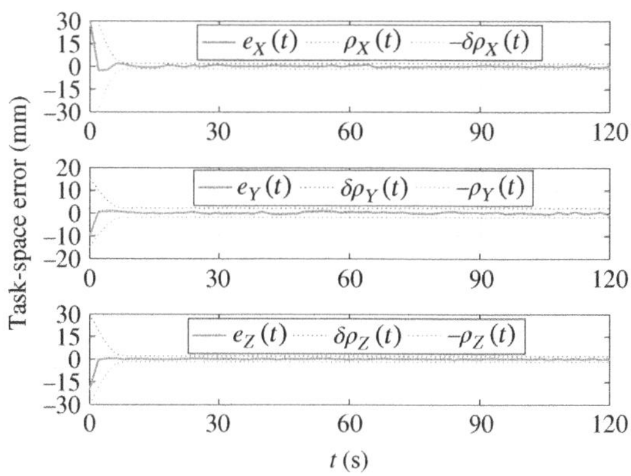

Figure 8.20 Experiment results of CCR tracking circular path.

of the BPAC method for continuum robots in achieving prescribed task-space performance.

8.7.2.2 Parallel Continuum Robot

Next, the PCR shown in Figure 8.21 was utilized to validate the BPAC method. The PCR prototype features 6 compliant legs, each of which can be actuated separately. The PCR was controlled to complete a tracking task. The initial position error between the robot's end-effector and the reference path was $\mathbf{e}(0) = [1, 1, 0]^{\mathrm{T}}$ mm. The parameters of the PPF were configured as $c = 0.5$, $\delta = 1$, $\boldsymbol{\rho}_0 = [10, 10, 10]^{\mathrm{T}}$ mm, and $\boldsymbol{\rho}_\infty = [1.5, 1.5, 1.5]^{\mathrm{T}}$ mm. The convergence parameter of the ZND models was assigned as $\lambda = 1$. The profiles of the reference path and the actual trajectory of the PCR are illustrated in Figure 8.22. The profiles of task-space

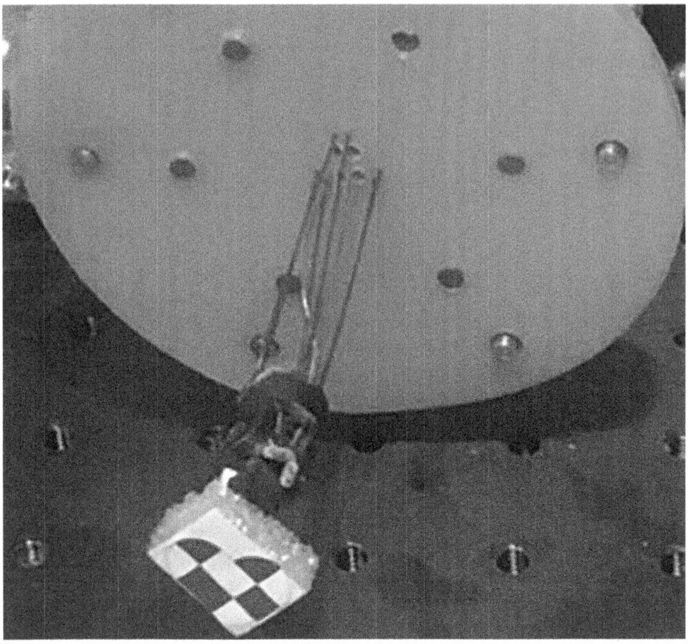

Figure 8.21 Illustration of physical PCR.

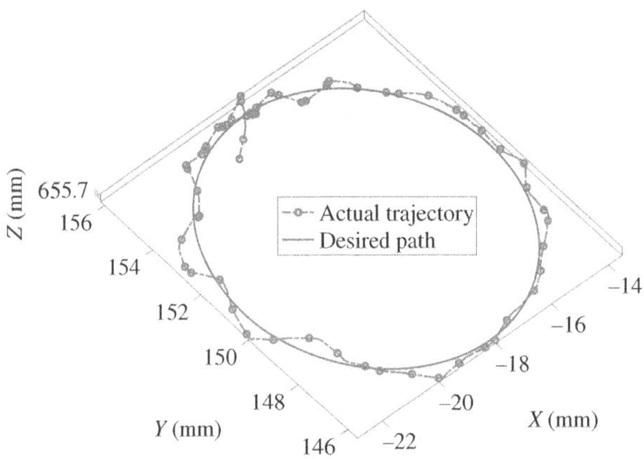

Figure 8.22 Trajectories of PCR performing circular path tracking task in experimental studies.

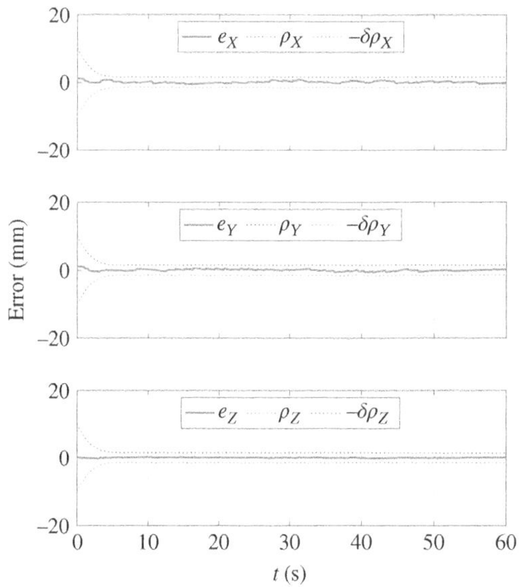

Figure 8.23 Experiment results of PCR tracking circular path.

errors are depicted in Figure 8.23, indicating that the tracking task was successfully completed by the PCR, with the task-space error remaining within specified boundaries.

8.7.2.3 Concentric-tube Robot

In this part, the CTR shown in Figure 8.24 was employed to validate the BPAC method. The CTR prototype consists of an inner tube and an outer tube, each of which has a specific curvature and can be rotated or translated. The PPF was applied to ensure that the robot's end-tip reaches the target position in $t_\mathrm{p} = 5$ s, with the residual task-space error being less than 1 mm. The initial position error between the robot's end-effector and the reference point was $\mathbf{e}(0) = [-10.2, 12.0, 0.9]^\mathrm{T}$ mm. The parameters of the PPF were configured as $c = 1$, $\delta = 1$, $\boldsymbol{\rho}_0 = [30, 30, 20]^\mathrm{T}$, and $\boldsymbol{\rho}_\infty = [1, 1, 1]^\mathrm{T}$. The convergence parameters of the ZND models were assigned as $\lambda_\mathrm{p} = 0.1$ and $\lambda_\mathrm{v} = 1$. The initial and final configurations of the CTR are depicted in Figure 8.24. The profiles of the task-space error are illustrated in Figure 8.25, indicating that the point-to-point task was successfully achieved by the CTR, with the task-space error remaining within the limits defined by the PPF. In summary, these experimental results illustrate the feasibility of the BPAC method for controlling the CTR.

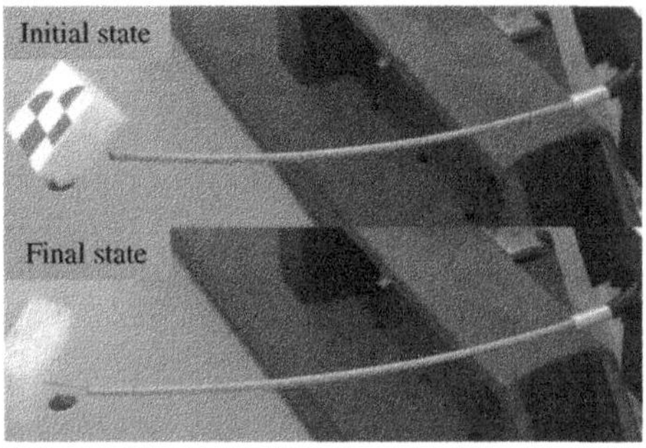

Figure 8.24 Illustration of physical CTR performing point-to-point task.

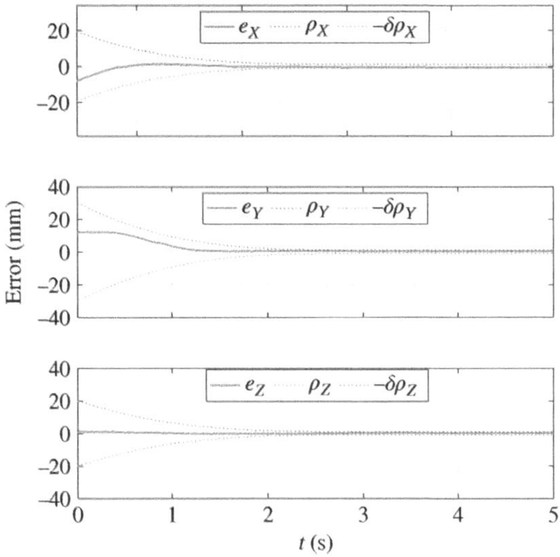

Figure 8.25 Experiment results of CTR performing point-to-point task.

8.7.2.4 Flexible Endoscope

In this part, the visual servoing experiment based on a flexible endoscope was conducted to further validate the BPAC method. Unlike previous experiments where the task space was Cartesian space, the task space of the flexible endoscope was the image space of the camera attached to the robot's end-tip. The objective was to achieve the desired image feature by adjusting the configuration of the endoscope. Various methods exist for representing image features, and for simplicity, we defined it as the pixel coordinate $\mathbf{p} = [x, y]^{\mathrm{T}} \in \mathbb{R}^2$ of a laser point in the image. The flexible endoscope was controlled using the BPAC method to ensure that the static laser point reaches the center of the image. The PPF was applied to ensure that the desired image feature is achieved in $t_p = 10$ s, with the residual task-space error being less than 15 pixels. The error between the initial image feature and the target image feature was $\mathbf{e}(0) = [-123, 121]^T$ pixels. The resolution of the image was 576×1024 pixels. The parameters of the PPF were set as $c = 0.3$, $\delta = 1$, $\boldsymbol{\rho}_0 = [236, 235]^T$ pixels, and $\boldsymbol{\rho}_\infty = [15, 15]^T$ pixels. The convergence parameters of the ZND models were set as $\lambda_p = 0.6$ and $\lambda_v = 1$. As illustrated in Figure 8.26, a target region was determined by the laser point, and the desired image feature was achieved successfully. It is evident from Figure 8.27 that the task-space error remained within the specified boundaries and converged to within 15 pixels before t_p, as expected.

8.7.2.5 Pneumatic Continuum Robot

In this section, the two-segment PneumCR shown in Figure 8.28 was adopted to validate the BPAC method.

The PneumCR was controlled to fulfill a point-to-point task. We controlled the robot end-effector to reach the target position in $t_p = 10$ s with the residual task-space error less than 10 mm. The initial position error between the robot end-effector and the reference point was $\mathbf{e}(0) = [80, 10, -20]^T$ mm. The parameters of the performance function are set as $c = 0.1$, $\delta = 1$, $\boldsymbol{\rho}_0 = [150, 100, 100]^T$ mm, and $\boldsymbol{\rho}_\infty = [10, 10, 10]^T$ mm. The convergence parameter of the ZND models was set as $\lambda = 1$. As depicted in Figure 8.28, the PneumCR reached the target point successfully. It can be seen from Figure 8.29 that the task-space error remained within the specified boundaries and converged to within 10 mm before t_p. These results have revealed the effectiveness of the BPAC method.

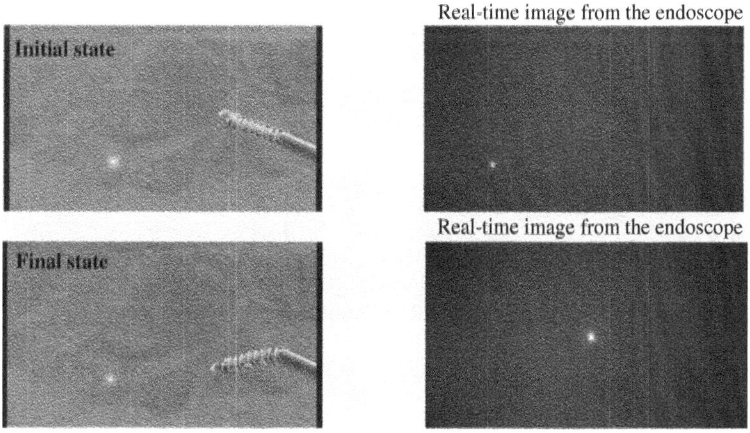

Figure 8.26 Illustration of physical flexible endoscope performing point-to-point task.

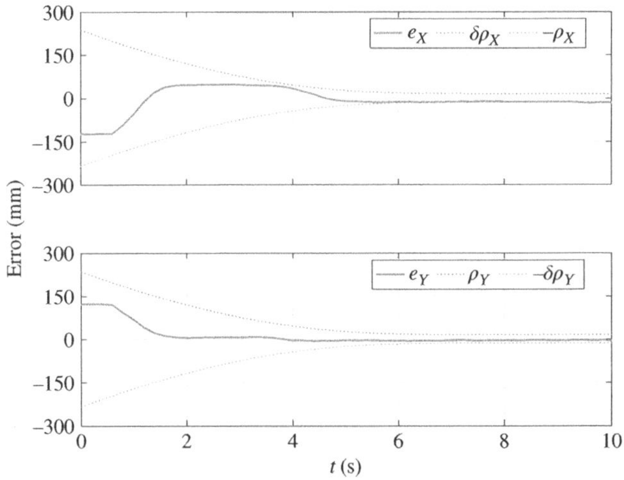

Figure 8.27 Experiment results of flexible endoscope performing point-to-point task.

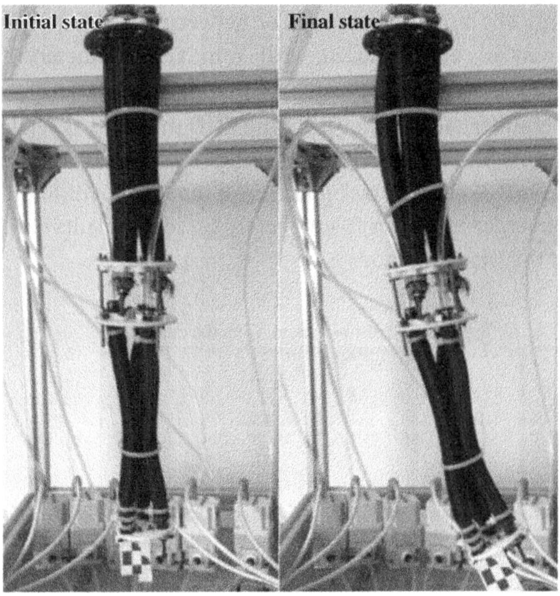

Figure 8.28 Illustration of physical PneumCR performing point-to-point task.

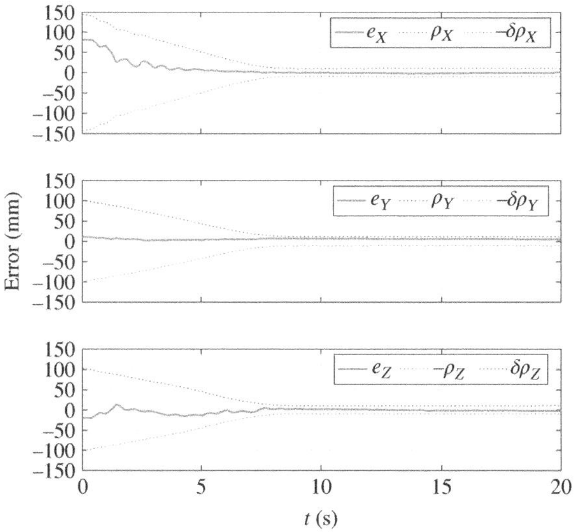

Figure 8.29 Experiment results of PneumCR performing point-to-point task.

8.8 Chapter Summary

This chapter introduces three control methods with prescribed task-space performance for HCRs. The control methods can prescribe the task-space performance in advance. The model-free nature of the methods provides the possibility of applying it to HCRs. Extensive simulative and experimental results on HCRs have confirmed that the kinematic motion control task of continuum robots can be achieved with prescribed performance. In summary, there are two main advantages of the control method compared to existing control methods. First, these methods can prescribe the task-space performance of continuum robots, thereby ensuring the efficiency and accuracy of end-effector tasks. Second, these methods can be easily and quickly applied to HCRs, which saves much effort for designing robot-specific or task-specific control methods.

Part VI

Enhancing Zeroing Neurodynamics Methods with Intelligent Algorithms

9

Fuzzy-enhanced Zeroing Neurodynamic Control Method

9.1 Introduction

Motion control and the improvement of positioning and tracking performance in continuum robots have been the focus of numerous studies. These efforts are generally classified into two main categories: model-based (MB) and model-less (ML) methods. MB approaches rely on the precise kinematic or dynamic models of robots. The first step in this methodology is to establish an accurate robot model, which is challenging when the robot's structure is complex or when model parameters are difficult to identify. Moreover, these approaches often operate under the strong assumption that the robot model is known accurately a priori. However, this assumption becomes problematic when the robot has been in operation for an extended period, causing the model parameters to deviate from their original values, or when the robot's fabrication lacks precision, which is often the case with rigid-continuum or soft robots. Additionally, the need to have prior knowledge of the robot model significantly reduces the generalizability of MB control methods. Each time a new robot model is introduced, the specific information about the model must be acquired beforehand, making the method less adaptable. The open-loop nature of MB methods also presents a significant challenge due to the accumulation of position errors, especially in long-term operations. These limitations of model-based approaches highlight the necessity of exploring model-less methods, which will be the primary focus of this chapter.

Fuzzy systems are promising in addressing error accumulation. Since the invention of fuzzy set theory, fuzzy control systems have been under continuous development for decades. It mimics human fuzzy inference and decision-making processes. The core component of a fuzzy control system is the fuzzy controller. The fuzzy controller, or the fuzzy logic controller, serves as a language-based controller, defined by a collection of "if-then" fuzzy logic control rules without exact mathematical models. Consequently, an advantage of the fuzzy control system is the ease with which fuzzy logic can be applied to solve complex and nonlinear problems, which has been widely applied in various areas, including robotics.

Due to their inherent suitability for hardware implementation, neurodynamic approaches have emerged as powerful tools for addressing both model-based and model-free control problems in continuum robots. For example, a method based on a zeroing neurodynamics (ZND) approach was proposed for solving the trajectory tracking problem of continuum robots [106]. Nevertheless, the conventional ZND scheme may suffer from error fluctuation during the trajectory tracking process. By adjusting control parameters of ZND, the error fluctuation can be suppressed. However, there is a lack of effective strategies for adjusting the parameters.

The combination of neurodynamic approaches and fuzzy logic represents an emerging trend in robot motion control. An adaptive fuzzy recurrent neural network has been proposed to address joint-drift issues in redundant robots [143]. While this method effectively addresses non-repetitive motion problems, it is predicated on the requirement that the kinematic model of the robot must be known a priori. As a further development, an adaptive fuzzy-type ZND model was developed to resolve time-variant quadratic programming (QP) problems [144], though it was not applied to robot control. This chapter proposes a fuzzy-enhanced neurodynamic scheme to solve model-less kinematic control for both redundant and hyper-redundant robots, marking a novel approach in this domain.

Fuzzy logic control technologies applied to hyper-redundant robots remain in the early stages, with only a few documented cases. The first attempt at kinematic control of continuum robots was made using a fuzzy-model-based approach for end-effector trajectory tracking in continuum robots [145]. However, the proposed fuzzy approach is challenged by high computational complexity, particularly when additional linearization points and rules are necessary to achieve a more accurate model. The T–S fuzzy technique was employed to design a robust feedback linearization observer aimed at enhancing the accuracy of fault estimation in surgical robots under uncertain conditions [146].

This chapter advances the field by introducing a novel dual fuzzy-enhanced neurodynamic (DFEN) scheme that addresses the trajectory tracking control problem of continuum robots without requiring prior knowledge of the exact robot models. Two fuzzy-logic-enhanced neurodynamic systems are specifically designed for kinematic control, effectively resolving the position error accumulation issue and achieving superior performance with strong robustness. The effectiveness and efficiency of the method are validated through simulations, which show improved tracking accuracy.

9.2 Method

In this section, the DFEN scheme is elaborated in three parts [147].

9.2.1 Fuzzy Adjustment Unit

For the control scheme, the fuzzy adjustment parameter ω_o plays a critical role in optimizing the performance of the dual neurodynamic scheme, which is enhanced by fuzzy logic techniques. The external environment and the position error of the end-effector are time-varying, and it is challenging to precisely describe the magnitude of the error. Therefore, a DFEN method is applied to enhance control accuracy. The adjustment parameter is generated through four key steps.

9.2.1.1 Fuzzification

For the input e and output ω_o of the fuzzy control system, the corresponding fuzzy input subset E and fuzzy output subset Ω are first defined. Next, a suitable membership function is selected. In this chapter, the triangular function and Gaussian function are chosen:

a. Triangular function

$$g(x) = \begin{cases} 0, & x \leq a \\ \dfrac{x-a}{b-a}, & a < x \leq b \\ \dfrac{c-x}{c-b}, & b < x \leq c \\ 0, & x \geq c \end{cases}, \tag{9.1}$$

b. Gaussian function

$$g(x) = \exp\left(-\frac{(x-c)^2}{2\sigma^2}\right), \tag{9.2}$$

where a, b, c, and σ are constants.

9.2.1.2 Fuzzy Rules

Fuzzy rules form the core of designing the fuzzy control system and are derived from the practical experience of experts. The fuzzy input and output subset includes zero (ZO), little small (LS), medium small (MS), little big (LB), and medium big (MB). Fuzzy rules are typically expressed as a collection of "if-then" rules. In this chapter, five fuzzy rules are designed for the fuzzy subsets E and Ω as follows:

- $S^{(1)}$: if $E = \text{ZO}$, then $\Omega = \text{ZO}$
- $S^{(2)}$: if $E = \text{LS}$, then $\Omega = \text{LS}$
- $S^{(3)}$: if $E = \text{MS}$, then $\Omega = \text{MS}$
- $S^{(4)}$: if $E = \text{LB}$, then $\Omega = \text{LB}$
- $S^{(5)}$: if $E = \text{MB}$, then $\Omega = \text{MB}$

Each fuzzy rule defines a specific aspect of the fuzzy system. As a result, the overall fuzzy relationship is formulated as follows:

$$S = S^{(1)} \cup S^{(2)} \cup S^{(3)} \cup S^{(4)} \cup S^{(5)}. \tag{9.3}$$

9.2.1.3 Fuzzy Inference Engine

After fuzzifying the input e, the fuzzy inference engine processes the input using the collection of fuzzy rules and completes the mapping from input to output. The process of fuzzy inference based on these fuzzy rules is described as follows:

$$\Omega = E \circ S, \tag{9.4}$$

where $\circ$ is the composition operator of the fuzzy relationship. The fuzzy output value is determined following the fuzzy inference process.

9.2.1.4 Defuzzification

The final step of the fuzzy control system is to convert the fuzzy output into a precise value, which is then used in the following control steps. In this case, the following centroid method is applied for defuzzification:

$$\omega_o = \frac{\int_\Omega \omega g_{E\circ S}(\omega)\mathrm{d}\omega}{\int_\Omega g_{E\circ S}(\omega)\mathrm{d}\omega}. \tag{9.5}$$

The key advantage of the centroid method is that it takes into account all elements within the fuzzy subset, leading to more accurate results. Using this fuzzy mechanism, the parameter ω_o is computed during the trajectory tracking task of the continuum robot.

9.2.2 Fuzzy-enhanced Controller

To begin with, the kinematic equation for a continuum robot is as follows:

$$\mathbf{f}(\mathbf{u}(t)) = \mathbf{p}(t), \tag{9.6}$$

where $\mathbf{u}(t) \in \mathbb{R}^n$ represents the actuation space vector at time t, and $\mathbf{p}(t) \in \mathbb{R}^m$ represents the end-effector position vector, with $\mathbf{f}(\cdot) : \mathbb{R}^n \rightarrow \mathbb{R}^m$ being a differentiable nonlinear function determined by the configuration of the continuum robot. Differentiating $\mathbf{p}(t)$ at time t gives the relationship between the end-effector velocity $\dot{\mathbf{p}}(t) \in \mathbb{R}^m$ and actuator velocity $\dot{\mathbf{u}}(t) \in \mathbb{R}^n$ as follows:

$$J(t)\dot{\mathbf{u}}(t) = \dot{\mathbf{p}}(t), \tag{9.7}$$

where $J(t) \in \mathbb{R}^{m\times n} = \partial\mathbf{f}(\mathbf{u}(t))/\partial\mathbf{u}(t)$ is the Jacobian matrix of the continuum robot.

The goal of trajectory tracking for continuum robots is to generate the actuation signal $\mathbf{u}(t) \in \mathbb{R}^n$ at time t based on the user-defined or desired path $\mathbf{p}_{\mathrm{d}}(t) \in \mathbb{R}^m$. To address the trajectory tracking problem, a vector-valued error function is defined as follows:

$$\mathbf{e}(t) = \mathbf{p}_{\mathrm{d}}(t) - \mathbf{p}(t). \tag{9.8}$$

To ensure that each element of the error function converges to zero, the following formula is developed using ZND and a fuzzy control strategy:

$$\dot{\mathbf{e}}(t) = -(\xi_1 + \omega_1)\mathbf{e}(t), \tag{9.9}$$

where $\xi_1 \in \mathbb{R}^+$ is a design parameter that adjusts the convergence rate, and ω_1 represents the fuzzy control value obtained from (9.5). The design formula (9.9) is then expanded to

$$\dot{\mathbf{p}}_{\mathrm{d}}(t) - J(t)\dot{\mathbf{u}}(t) = -(\xi_1 + \omega_1)(\mathbf{p}_{\mathrm{d}}(t) - \mathbf{p}(t)). \tag{9.10}$$

From (9.10), the control input signal at velocity level is derived as follows:

$$\dot{\mathbf{u}}(t) = J^{\dagger}(t)(\dot{\mathbf{p}}_{\mathrm{d}}(t) + (\xi_1 + \omega_1)(\mathbf{p}_{\mathrm{d}}(t) - \mathbf{p}(t))). \tag{9.11}$$

9.2.3 Fuzzy-enhanced Jacobian Estimator

For the sake of the adaption of the Jacobian matrix, a vector-valued error function is defined as follows:

$$\mathbf{e}_J(t) = \dot{\mathbf{p}}(t) - J(t)\dot{\mathbf{u}}(t). \tag{9.12}$$

Then, inspired by the ZND method and the fuzzy logic, the following formula is designed:

$$\dot{\mathbf{e}}_J(t) = -(\xi_2 + \omega_2)\mathbf{e}_J(t), \tag{9.13}$$

where $\xi_2 \in \mathbb{R}^+$ is a design parameter that adjusts the convergence rate, and ω_2 represents the fuzzy control value also obtained through (9.5). Using (9.12), (9.13) is then expanded to

$$\ddot{\mathbf{p}}(t) - \dot{J}(t)\dot{\mathbf{u}}(t) - J(t)\ddot{\mathbf{u}}(t) = -(\xi_2 + \omega_2)(\dot{\mathbf{p}}(t) - J(t)\dot{\mathbf{u}}(t)). \tag{9.14}$$

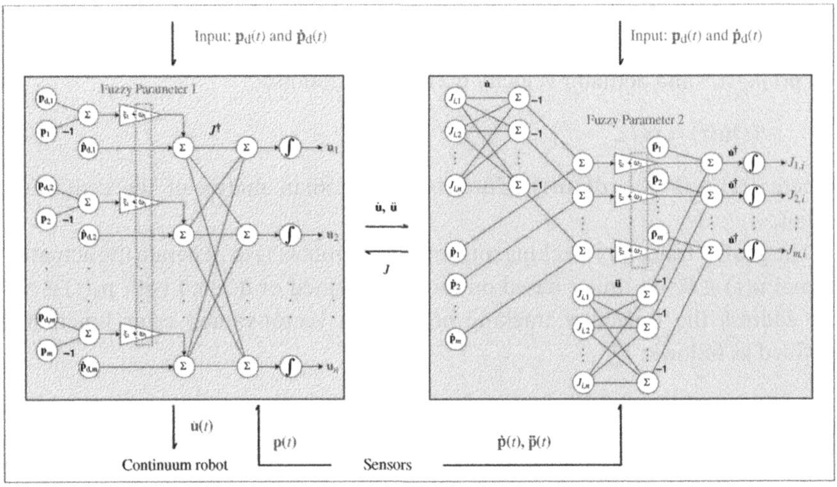

Figure 9.1 Structure of DFEN method.

Next, the following solution is derived:

$$\dot{J}(t) = (\ddot{\mathbf{p}}(t) - J(t)\ddot{\mathbf{u}}(t) + (\xi_2 + \omega_2)(\dot{\mathbf{p}}(t) - J(t)\dot{\mathbf{u}}(t)))\dot{\mathbf{u}}^{\dagger}(t), \tag{9.15}$$

where $\dot{\mathbf{u}}^{\dagger}(t)$ represents the pseudo-inverse of the actuation velocity vector. The parameters ξ_1, ω_1, ξ_2, and ω_2 in the differential equations (9.11) and (9.15) are designed using ZND and the fuzzy control strategy. Finally, formulas (9.11) and (9.15) represent the dynamic equations of the DFEN system.

Solving the differential equations (9.11) and (9.15) provides the desired $\mathbf{u}(t)$ and $J(t)$, where the Jacobian matrix is approximated in a data-driven manner rather than being derived from the kinematic model. For a clearer understanding of the method, the structure of the DFEN scheme is illustrated in Figure 9.1. Using the inputs and real-time sensor outputs, the end-effector's position $\mathbf{p}(t)$, velocity $\dot{\mathbf{p}}(t)$, and acceleration $\ddot{\mathbf{p}}(t)$ can be obtained as feedback for the closed-loop control system. This allows the trajectory tracking problem to be solved while adaptively learning the unknown Jacobian matrix.

9.3 Verification

In this section, a two-segment continuum robot is used to verify the DFEN control method. As a comparison, the basic ZND method without fuzzy enhancement is also adopted to fulfill the same task. As shown in Figure 9.2, Gaussian and

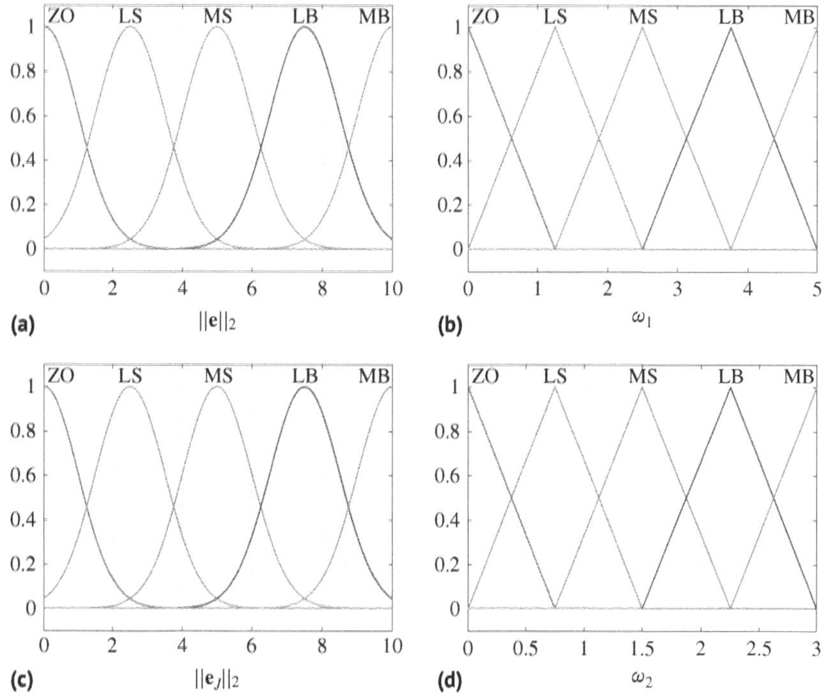

Figure 9.2 Membership functions of fuzzy control system for continuum robot: (a) membership function of input $\| \mathbf{e} \|_2$; (b) membership function of output ω_1; (c) membership function of input $\| \mathbf{e}_J \|_2$; (d) membership function of output ω_2.

triangular membership functions are applied to the inputs ($\| \mathbf{e} \|_2$ and $\| \mathbf{e}_J \|_2$) and the corresponding outputs (ω_1 and ω_2), respectively. The centroid defuzzification method is used. The trajectory task duration is set to $T_d = 20$ s. The initial actuation state of the continuum robot is $\mathbf{u}(0) = [3.0, 0, 0, 3.0, 0, 0]^T$ mm. The initial value of the Jacobian matrix is determined following the motor babbling strategy in the previous chapters. Specifically, the initial value of the Jacobian matrix for the adaptation process is set as follows:

$$J(0) = \begin{bmatrix} 13.8563 & -0.0006 & 0.1489 & 4.37600 & -0.0006 & 0.4333 \\ 0 & 14.4445 & 0 & 0 & 4.96260 & 0 \\ -5.3305 & -0.0064 & 0.9851 & -2.3927 & -0.0015 & 0.8971 \end{bmatrix}.$$

For the trajectory tracking task in this section, a rhodonea path is considered. During execution, the x-, y-, and z-axis of the desired end-effector path $\mathbf{p}_{\rm d}(t)$ are set as follows:

$$\phi = 2\pi \sin\left(\frac{0.5\pi t}{T_{\rm d}}\right) \sin\left(\frac{0.5\pi t}{T_{\rm d}}\right)$$

$$p_{{\rm d},x}(t) = \iota \cos(2\phi)\cos(\phi) + 58.2215 - \iota$$

$$p_{{\rm d},y}(t) = \iota \cos\left(\frac{\pi}{6}\right)\cos(2\phi)\sin(\phi)$$

$$p_{{\rm d},z}(t) = \iota \sin\left(\frac{\pi}{6}\right)\cos(2\phi)\sin(\phi) + 188.2142,$$

where the geometry parameter is set as $\iota = 35$ mm.

The corresponding simulation results of tracking the rhodonea path are illustrated in Figure 9.3, which shows that the actual trajectory of the end-effector of the continuum robot is sufficiently close to the desired circular path. Figure 9.4 illustrates the end-effector position errors $\mathbf{e} = [e_x, e_y, e_z]^{\rm T}$. The method effectively resolves the issue of position error accumulation. The lengths of the continuum robot segments vary smoothly and remain stable throughout the tracking process.

The ZND method was introduced as an alternative tracking control strategy to solve the tracking problem. It is described as follows:

$$\dot{\mathbf{u}}(t) = J^{\dagger}(t)(\dot{\mathbf{p}}_{\rm d}(t) + \gamma(\mathbf{p}_{\rm d}(t) - \mathbf{p}(t))), \tag{9.16}$$

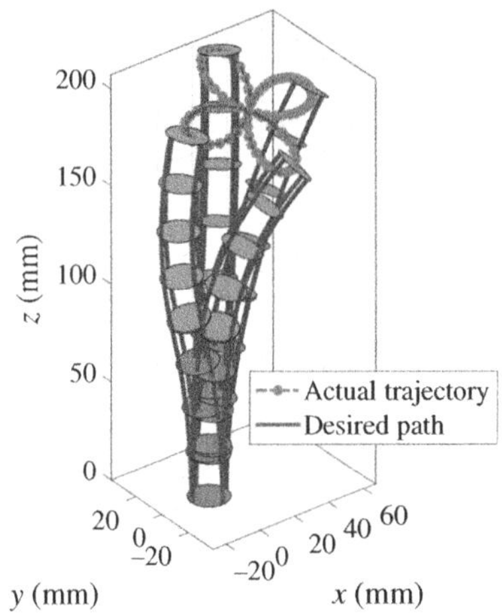

Figure 9.3 Trajectory tracking process with aid of DFEN method.

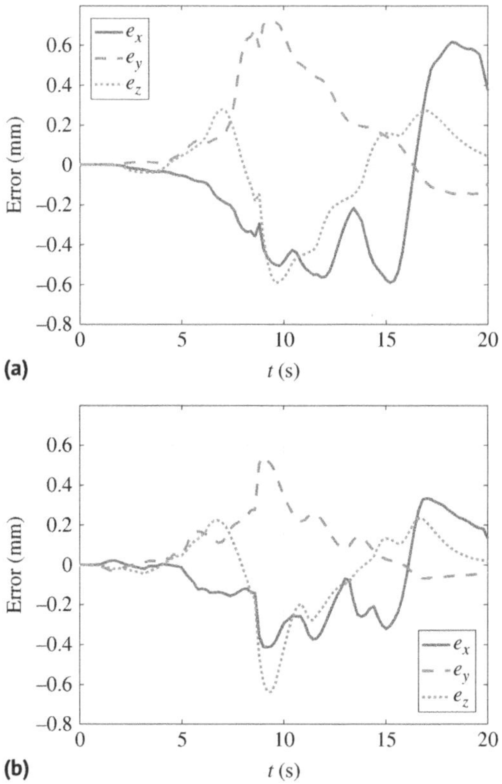

Figure 9.4 Trajectory tracking errors of different methods: (a) errors of ZND; (b) errors of DFEN.

where $\gamma \in \mathbb{R}^n$ is a design parameter used to scale the convergence rate. The comparative simulation was conducted on the same continuum robot, and all simulation conditions were kept the same. The comparative position errors synthesized by the ZND method are depicted in Figure 9.4a. The maximal absolute values of the end-effector position error in this case exceed those observed in the simulation of the DFEN method. This is further validated in Table 9.1 through the root-mean-square error (RMSE) values.

In summary, compared to the ZND method, the end-effector position errors of the DFEN method converge to a smaller range, illustrating that the DFEN scheme in this chapter not only achieves greater accuracy but also shows superior capability in handling position error accumulation. Furthermore, the simulations illustrate that the DFEN scheme effectively addresses the tracking control problem of the continuum robot with unknown model information.

Table 9.1 RMSE for different methods when tracking rhodonea path.

Method	Model	RMSE (mm)
ZND	model-free	0.5271
DFEN	model-free	0.3331

9.4 Chapter Summary

This chapter introduces a novel DFEN method based on ZND to solve the trajectory tracking problem of continuum robots with enhanced performance. With a self-adaptive convergence rate based on errors, the method effectively completes the tracking task and reduces the task-space error. Comparison with the ZND method has confirmed the merits of the DFEN method.

10

Spiking Neural Network-enhanced Zeroing Neurodynamic Control Method

10.1 Introduction

Neural networks are the primary driving force behind the current advancements in artificial intelligence and have undergone multiple stages of evolution over the past century. However, artificial neural networks (ANNs) are biologically inaccurate and fail to adequately simulate the operational mechanisms of the real biological cerebellum. Inspired by advances in brain science, spiking neural networks (SNNs) are recognized as a new generation of neural networks characterized by rich neurodynamic properties, various coding mechanisms, and event-driven, low energy consumption features [148]. Maass [149] stated that, compared to previous neural networks, SNNs utilize a smaller number of neurons and thus exhibit greater computational power. Even in the worst case, SNNs retain equivalent computational power. Brain-inspired computing, grounded in the findings of brain science, remains in the early stages of rapid development. Although SNNs incorporate training algorithms such as spike-timing dependent plasticity and ANN-to-SNN, they still face challenges related to encoding the spiking signal and the absence of a uniform, effective training method due to the discontinuity of spiking events. Drawing inspiration from the structure of the mammalian central nervous system, Maass et al. [150] proposed a reservoir computational system based on spiking neurons, termed the liquid state machine (LSM). This chapter aims to augment the zeroing neurodynamics (ZND) approach using LSM.

In recent years, several automation control algorithms have been introduced into medical surgery [151]. By considering vision as an integrated sensor of the endoscope, robotic endoscope automation can establish a control system utilizing the visual signals from the endoscope camera. Visual servoing control enables the robotic endoscope to automatically track the target of interest, eliminating the need for the surgeon to exert additional effort in stabilizing the target at the center of the image. Due to their superior safety and flexibility compared to rigid structures, soft robots have been extensively used in visual servoing tasks [152]. During minimally invasive surgery (MIS), surgical instruments are inserted into the patient's body through a small incision in the abdominal wall. The small incision imposes constraints on the inserted robot end-effector, commonly referred to as the remote center of motion (RCM) constraints [153]. RCM constraints necessitate that the surgical manipulator performs its task while minimizing the movement around the incision to prevent enlarging the wound. However, maintaining the RCM constraints during surgical operations presents a significant challenge.

Numerous control strategies have considered the RCM constraints [154–156]. Model-based control methods addressing RCM constraints have made significant advancements. However, in practical applications, some kinematic parameters may be unavailable, or the calibrated parameters may not align with the actual values, rendering these model-based control methods less feasible. Recently, model-free control methods have been proposed for handling robotic arms with unknown parameters [157, 158]. Motivated by the advancements in artificial intelligence and the fact that humans possess powerful learning and precise motor capabilities, there has been an increasing search for a model-free, cerebellum-inspired scheme.

Humans can accurately perform a variety of complex tasks despite their intricate skeletal structures and nondeterministic sensory feedback delays. Drawing inspiration from cerebellar motor control biology, this chapter develops a control scheme based on the SNN and ZND approach for a rigid-flexible hybrid robotic endoscope, which includes a redundant robotic manipulator for tracking tasks with RCM constraints and a flexible surgical endoscope for visual servoing tasks [159].

10.2 Preliminary

This section discusses the visual servoing control of the flexible endoscope and the kinematics problem of redundant manipulators with RCM constraints.

10.2.1 Remote Center of Motion Constraints

During the surgical procedure, the robot's shaft must pass through the initial insertion point ($\mathbf{p}_{\text{trocar}} \in \mathbb{R}^3$) without exhibiting excessive movement due to the robot arm's motion, as shown in Figure 10.1b.

For an m-DoF robot manipulator with a shaft, the mappings from its joint space to the Cartesian coordinate of the robot manipulator's endpoint $\mathbf{p}_m \in \mathbb{R}^3$, the shaft

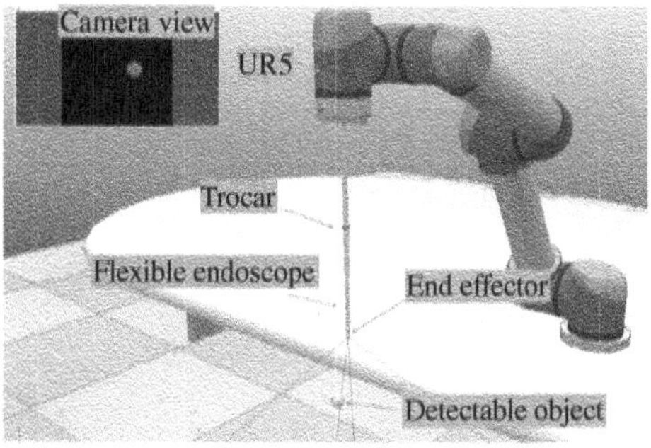

(a)

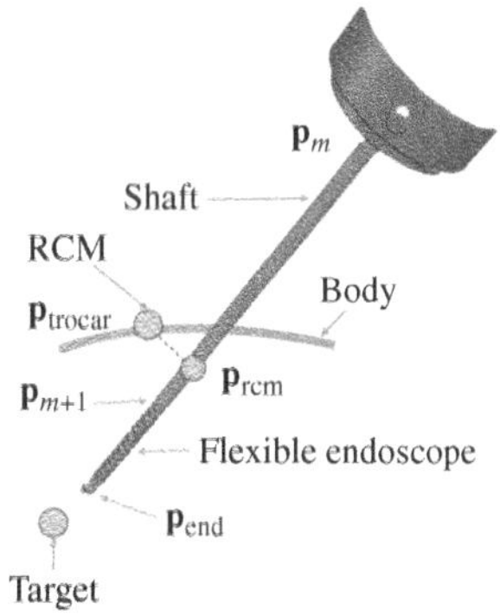

(b)

Figure 10.1 Robotic endoscope system: (a) rigid-flexible hybrid robot; (b) remote center of motion.

tip $\mathbf{p}_{m+1} \in \mathbb{R}^3$, and the flexible endoscope's endpoint $\mathbf{p}_{\text{end}} \in \mathbb{R}^3$ are expressed by the following equations:

$$\begin{aligned} \mathbf{p}_m &= \mathbf{f}_m(\mathbf{u}), \\ \mathbf{p}_{m+1} &= \mathbf{f}_{m+1}(\mathbf{u}), \\ \mathbf{p}_{\text{end}} &= \mathbf{f}_{\text{end}}(\mathbf{u}). \end{aligned} \tag{10.1}$$

To satisfy the RCM constraints, $\mathbf{p}_{\text{trocar}}$ should be as close as possible to the straight line $\overrightarrow{\mathbf{p}_m\mathbf{p}_{m+1}}$ connecting $\mathbf{p}_m$ and $\mathbf{p}_{m+1}$. The RCM point $\mathbf{p}_{\text{rcm}}$ is defined as the projection of $\mathbf{p}_{\text{trocar}}$ onto the straight line $\overrightarrow{\mathbf{p}_m\mathbf{p}_{m+1}}$ as follows:

$$\begin{aligned} k &= -((x_m - x_{\text{trocar}})(x_{m+1} - x_m) + (y_m - y_{\text{trocar}})(y_{m+1} - y_m) \\ &\quad + (z_m - z_{\text{trocar}})(z_{m+1} - z_m)), \\ x_{\text{rcm}} &= \frac{k(x_{m+1} - x_m)}{l_1^2} + x_m, \\ y_{\text{rcm}} &= \frac{k(y_{m+1} - y_m)}{l_1^2} + y_m, \\ z_{\text{rcm}} &= \frac{k(z_{m+1} - z_m)}{l_1^2} + z_m, \end{aligned} \tag{10.2}$$

where $\mathbf{p}_{\mathrm{rcm}} = [x_{\mathrm{rcm}}, y_{\mathrm{rcm}}, z_{\mathrm{rcm}}]^{\mathrm{T}}$ and $l_1 = \|\mathbf{p}_{m+1} - \mathbf{p}_m\|_2$ is the length of $\overrightarrow{\mathbf{p}_m\mathbf{p}_{m+1}}$. Then the error of RCM is represented as

$$\varepsilon_{\mathrm{r}} = \mathbf{p}_{\mathrm{rcm}} - \mathbf{p}_{\mathrm{trocar}}. \tag{10.3}$$

The point $\mathbf{p}_{\mathrm{rcm}}$ should satisfy the following equation:

$$J_{\mathrm{r}}(t)\dot{\mathbf{u}}(t) = \dot{\mathbf{p}}_{\mathrm{rcm}}(t), \tag{10.4}$$

where $J_{\mathrm{r}} \in \mathbb{R}^{3\times m}$ is the Jacobian matrix corresponding to the RCM point.

10.2.2 Flexible Endoscope Kinematics for Visual Servoing

As illustrated in Figure 10.2, the detectable object $\mathbf{p} = [x, y, z]^{\mathrm{T}}$ is projected onto the point $\mathbf{p}_{\mathrm{v}} = [u, v]^{\mathrm{T}}$ on the image plane. The visual servoing task involves controlling the flexible endoscope so that the projected point $\mathbf{p}_{\mathrm{v}}$ gradually approaches the target point $\mathbf{p}_{\mathrm{u}}$ on the image plane. The error of the visual servoing task is then expressed as follows:

$$\varepsilon_{\mathrm{v}} = \mathbf{p}_{\mathrm{v}} - \mathbf{p}_{\mathrm{u}}. \tag{10.5}$$

The velocity-level kinematic equation for the velocity of the point $\dot{\mathbf{p}}_{\mathrm{v}} \in \mathbb{R}^2$ on the image plane and the velocity of the endoscope $\dot{\boldsymbol{\vartheta}} \in \mathbb{R}^h$ are shown as follows:

$$J_{\mathrm{v}}\dot{\boldsymbol{\vartheta}} = \dot{\mathbf{p}}_{\mathrm{v}}, \tag{10.6}$$

where $J_{\mathrm{v}} \in \mathbb{R}^{2\times h}$ is the Jacobian matrix of the robotic endoscope system.

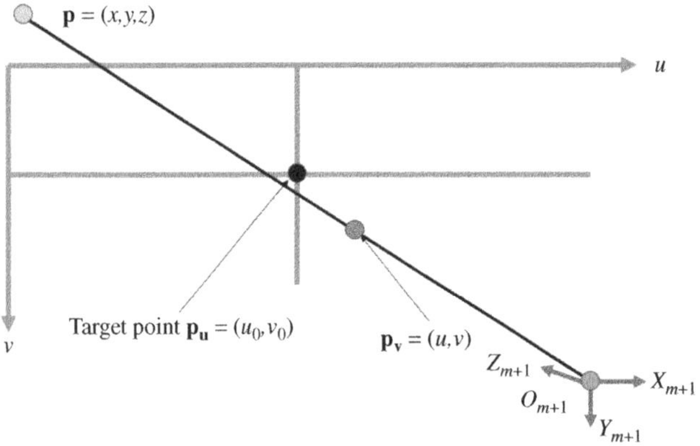

Figure 10.2 Endoscope camera model.

10.3 LSM-based Control Scheme

This section introduces the robotic manipulator performing a tracking or regulation task (e.g., observing a lesion or moving the end-effector to a specific position) under RCM constraints. Once the targeting area is reached, the flexible endoscope performs the visual servoing task to keep the scene of interest (e.g., an organ or tissue) within the field of view during the operation. These two steps could be repeated after the entire operation is completed.

10.3.1 LSM-based Cerebellum Model

LSM is a reservoir computing system where multiple input signals are injected into a nonlinear dynamic system using spiking neurons. The resulting spiking sequences are collected and processed in an external layer known as the readout [150]. The functional roles of the liquid state generator and readout neurons correspond to the granular layer and Purkinje cells (PCs) in the cerebellar cortex, respectively. As illustrated in Figure 10.3, the cerebellum model based on the LSM includes mossy fibers (MFs), granular cells (GCs), parallel fibers (PFs), climbing fibers (CFs), and PCs. The input signal $\mathbf{x}(t) \in \mathbb{R}^K$ of the cerebellum model is transmitted to the granular layer via the MFs at each simulation time step (e.g., 2 ms). No standardized method exists for encoding sensor data into spiking events. Common methods include population coding and probabilistic spike sampling. The MFs are divided into K MF neuron subgroups, each corresponding to an input

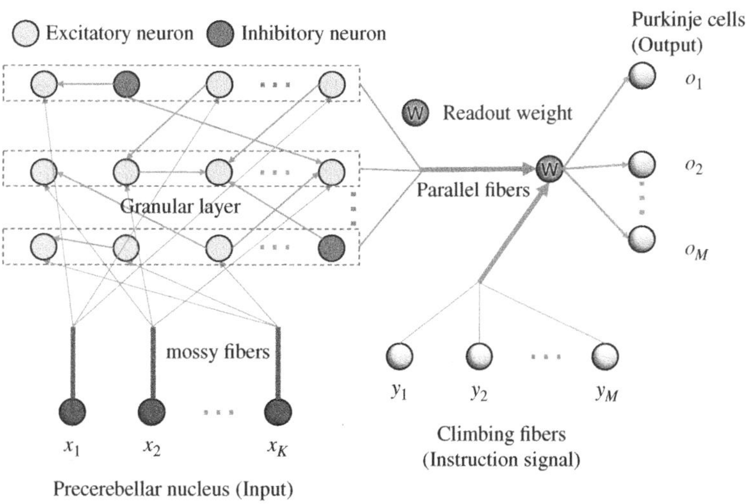

Figure 10.3 Diagram of cerebellum network model based on LSM.

signal, and are responsible for converting sensorimotor information into spiking events using the simplified population coding method before transmitting it to the granular layer.

The granular layer consists of GCs modeled as leaky integrate-and-fire (LIF) neurons [160], which are connected recurrently. The granular layer acts as a reservoir and is responsible for retaining and presenting coded sensorimotor information from the MFs. Several spiking neuron models have been proposed. In this chapter, the LIF neuron with exponential synaptic dynamics is used to closely approximate the real biological model and reduce computational overhead. The model is defined by a set of equations as follows:

$$\frac{\mathrm{d}\upsilon}{\mathrm{d}t} = \frac{i_{\mathrm{e}}(t) + i_{\mathrm{i}}(t) + i_{\mathrm{offset}} + i_{\mathrm{noise}}}{c_{\mathrm{m}}} + \frac{\upsilon_{\mathrm{rest}} - \upsilon}{\tau_m}, \tag{10.7}$$

$$\frac{\mathrm{d}i_{\mathrm{e}}}{\mathrm{d}t} = -\frac{i_{\mathrm{e}}}{\tau_{\mathrm{syn_e}}}, \tag{10.8}$$

$$\frac{\mathrm{d}i_{\mathrm{i}}}{\mathrm{d}t} = -\frac{i_{\mathrm{i}}}{\tau_{\mathrm{syn_i}}}, \tag{10.9}$$

where υ is the membrane potential, υ_{rest} is the resting membrane potential, τ_m is the membrane time constant, c_{m} is the membrane capacitance, i_{offset} and i_{noise} can be seen as coming from different noise sources, $\tau_{\mathrm{syn_e}}$ is the decay time of the excitatory synaptic current, while $\tau_{\mathrm{syn_i}}$ is the decay time of the inhibitory one. When the membrane potential of an LIF neuron is greater than the membrane threshold ($\upsilon_{\mathrm{threshold}} = 15$ mV), the neuron generates a spike.

The granular layer consists of N LIF neurons arranged in a 3D structure, with 80% excitatory (E) neurons and 20% inhibitory (I) neurons. The granular layer is divided into K GCs subgroups, each corresponding to an input signal. Excitatory neurons in each GCs subgroup are fully connected to the corresponding subgroup in the MFs. The probability of generating connections between GCs depends on the type and distance of neurons as follows:

$$P_{i,j} = c \exp\left(-\frac{D_{i,j}^2}{\lambda^2}\right), \tag{10.10}$$

where the value of λ is 1.2 and c is determined by the type of neurons: 0.3 (EE), 0.2 (EI), 0.4 (IE), and 0.1 (II). EE, EI, IE, and II denote the connections of different types of neurons. EE represents the connection between two different excitatory neurons. EI represents the connection from excitatory to inhibitory neurons. IE denotes the connection from inhibitory to excitatory neurons, and II denotes the connection between two different inhibitory neurons.

The spiking events of the GCs are relayed by means of PFs to PCs. The signal $\mathbf{z}(t) \in \mathbb{R}^N$ of the PFs is described as follows:

$$z_i(t) = \begin{cases} 0, & \text{if } v_i(t) < v_{\text{threshold}} \\ 1, & \text{else} \end{cases}, \tag{10.11}$$

where $v_i(t)$ represents the voltage of the ith GC neuron at time t, and $v_{\text{threshold}}$ denotes the threshold voltage.

The output layer consists of the PCs, which can handle the high-dimensional space of the granular layer in a simple way. The output $\mathbf{o}(t) \in \mathbb{R}^M$ of the PCs is modeled as a weighted sum of the signal $\mathbf{z}(t)$ of the PFs:

$$\mathbf{o}(t) = W_{\text{out}}(t)\mathbf{z}(t), \tag{10.12}$$

where $W_{\text{out}}(t) \in \mathbb{R}^{M \times N}$ is the output weight. When the LSM scheme is used, linear regression is usually implemented to train the output weight of the connected PFs and PCs. In this chapter, the ordinary least square (OLS) is used to solve the following problem:

$$\min_{W_{\text{out}}} \|W_{\text{out}}Z - Y\|_2^2, \tag{10.13}$$

where Z is usually composed of a sequence of the signal $\mathbf{z}$ of the PFs, and Y is a set of instruction signals from CFs to be learned.

It is observed from Figure 10.1 that the motion of the endoscope itself does not involve the RCM constraints. Therefore, unlike Li et al. [161], which operates both the robotic arm and the endoscope to accomplish the visual servoing task with RCM constraints, this chapter decouples the control problem into operating the robotic manipulator to accomplish the tracking task with RCM constraints and operating the endoscope to accomplish the visual servoing task. For the redundant manipulator with k joints and the flexible endoscope with h joints, two cerebellum-like spiking models based on LSM are created as the corresponding controllers. The parameters of (10.7)–(10.9) are chosen as follows: $i_{\text{noise}} \sim U(13.5, 14.5)$ nA, $i_{\text{noise}} \sim \mathcal{N}(0, 0.01)$ nA, $c_{\text{m}} = 30$ nF, $v_{\text{rest}} = 0$ mV, $\tau_m = 30$ ms, $\tau_{\text{syn}_\text{e}} = 3$ ms, and $\tau_{\text{syn}_\text{i}} = 6$ ms.

In the tracking task, the robotic manipulator's joint angle $\theta(t) \in \mathbb{R}^k$ as the input signal $\mathbf{x}(t) = \theta(t) \in \mathbb{R}^k$ is fed to cerebellum model I. Next, the cerebellum model I generates the output signal

$$\mathbf{o}(t) = W_{\text{out}}(t)\mathbf{z}(t) \in \mathbb{R}^k \tag{10.14}$$

as joint velocity $\dot{\theta}_{\text{out}}(t)$, where $\mathbf{z}(t) \in \mathbb{R}^N$ is the spiking sequence of the liquid layer at time t and $N = 100k$ is the number of all LIF neurons in the granular layer, which means that each GCs subgroup has 100 LIF neurons. Correspondingly, in the visual servoing task, the input signal $\mathbf{x}(t) = \vartheta(t) \in \mathbb{R}^h$ (i.e., the joint angle of the endoscope) is fed to the cerebellum model II and the model outputs the joint velocity of the endoscope $\dot{\vartheta}_{\text{out}}(t) = \mathbf{o}(t) \in \mathbb{R}^h$. The number of all GCs N in the cerebellum model II is equal to $100h$.

However, many previous control methods based on LSMs demanded pretraining the output weight parameters W_{out}, which requires to model the robot manipulator and collects a certain amount of data in advance. Since the training of the output weight is based on supervised learning algorithms, in the following, the cerebellum network is combined with ZND to establish a model-free, pretraining-free, and online learning approach.

10.3.2 Tracking Control with RCM Constraints

According to (10.4), we obtain the two desired joint velocities for the tracking task and for the RCM constraints, respectively, as follows:

$$\begin{aligned} \dot{\theta}_{\text{t}}(t) &= J_{\text{t}}^{\dagger}(t)\dot{\varepsilon}_{\text{t}}(t), \\ \dot{\theta}_{\text{r}}(t) &= J_{\text{r}}^{\dagger}(t)\dot{\varepsilon}_{\text{r}}(t), \end{aligned} \tag{10.15}$$

where $(\cdot)^{\dagger}$ is the pseudo-inverse operation, $\dot{\varepsilon}_{\text{t}}(t)$ is the derivative of $\varepsilon_{\text{t}}(t)$, and $\dot{\varepsilon}_{\text{r}}(t)$ is the derivative of $\varepsilon_{\text{r}}(t)$. How to handle these two desired joint velocities appropriately is a key point, since the two tasks affect each other in most cases. Inspired by Siciliano and Slotine [162], the following joint velocity solution is designed:

$$\dot{\theta}_{\text{d}}(t) = \dot{\theta}_{\text{t}}(t) + \widetilde{J}^{\dagger}(t)(\dot{\varepsilon}_{\text{r}}(t) - J_{\text{r}}(t)\dot{\theta}_{\text{t}}(t)), \tag{10.16}$$

where

$$\widetilde{J} = J_{\text{r}}(I - J_{\text{t}}^{\dagger}J_{\text{t}}) \tag{10.17}$$

with I being an identity matrix. However, matrices J_{t} and J_{r} require accurate prior modeling of the robot model, and the Jacobian matrix J_{r} also involves the specific insertion point $\mathbf{p}_{\text{trocar}}$, which makes modeling more difficult. Therefore, two ZNDs are introduced to estimate the Jacobian matrices $\hat{J}_{\text{t}}$ and $\hat{J}_{\text{r}}$, respectively. A vector-valued error function is defined to measure the error when estimating the Jacobian matrix $\hat{J}_{\text{t}}$ used in (10.15):

$$\mathbf{e}_{\text{t}}(t) = \dot{\mathbf{p}}_{\text{a}}(t) - \hat{J}_{\text{t}}\dot{\theta}(t). \tag{10.18}$$

The following equation is defined according to the ZND design principle:

$$\dot{\mathbf{e}}_{\text{t}}(t) = -\phi\mathbf{e}_{\text{t}}(t), \tag{10.19}$$

where ϕ is a positive parameter for the convergence adjustment and $\dot{\mathbf{e}}_\mathrm{t}(t)$ is the derivative of $\mathbf{e}_\mathrm{t}(t)$ with respect to time. Substituting (10.18) into (10.19) yields

$$\dot{\hat{J}}_\mathrm{t}(t) = (\ddot{\mathbf{p}}_\mathrm{a}(t) - \hat{J}_\mathrm{t}(t)\ddot{\theta}(t) + \phi(\dot{\mathbf{p}}_\mathrm{a}(t) - \hat{J}_\mathrm{t}(t)\dot{\theta}(t)))\dot{\theta}^{\dagger}(t), \quad (10.20)$$

where $\dot{\hat{J}}_\mathrm{t}(t)$ represents the derivative of $\hat{J}_\mathrm{t}(t)$ with respect to t, $\ddot{\mathbf{p}}_\mathrm{a}(t)$ is the derivative of $\dot{\mathbf{p}}_\mathrm{a}(t)$, $\ddot{\theta}(t)$ is the derivative of $\dot{\theta}(t)$, and $\dot{\theta}^{\dagger}(t)$ represents the pseudo-inverse of $\dot{\theta}(t)$. The equation for $\hat{J}_\mathrm{r}$ is obtained similarly:

$$\dot{\hat{J}}_\mathrm{r}(t) = (\ddot{\mathbf{p}}_\mathrm{rcm}(t) - \hat{J}_\mathrm{r}(t)\ddot{\theta}(t) + \nu(\dot{\mathbf{p}}_\mathrm{rcm}(t) - \hat{J}_\mathrm{r}(t)\dot{\theta}(t)))\dot{\theta}^{\dagger}(t), \quad (10.21)$$

where $\dot{\hat{J}}_\mathrm{r}(t)$ represents the derivative of $\hat{J}_\mathrm{r}(t)$ with respect to t, $\ddot{\mathbf{p}}_\mathrm{rcm}(t)$ is the derivative of $\dot{\mathbf{p}}_\mathrm{rcm}(t)$, and $\nu \in \mathbb{R}^+$ is another design parameter of ZND. Consequently, the training signal $\dot{\hat{\theta}}_\mathrm{d}(t)$ is obtained to calculate the output weight W_out in the cerebellum model I:

$$\begin{aligned}
\dot{\hat{\theta}}_\mathrm{d}(t) &= \dot{\hat{\theta}}_\mathrm{t}(t) + \widetilde{J}^{\dagger}(t)(\dot{\varepsilon}_\mathrm{r}(t) - \hat{J}_\mathrm{r}(t)\dot{\hat{\theta}}_\mathrm{t}(t)),\\
\dot{\hat{\theta}}_\mathrm{t}(t) &= \hat{J}_\mathrm{t}^{\dagger}(t)\dot{\varepsilon}_\mathrm{t}(t), \quad (10.22)\\
\widetilde{J} &= \hat{J}_\mathrm{r}(I - \hat{J}_\mathrm{t}^{\dagger}\hat{J}_\mathrm{t}).
\end{aligned}$$

In order to achieve real-time training of the output weight $W_\mathrm{out}(t+1)$, this chapter takes the spiking sequences of the PFs at the previous g moments to form $Z_g \in \mathbb{R}^{N\times g}$ and the corresponding instruction signals calculated by (10.22) to form $Y_k = [\dot{\hat{\theta}}_\mathrm{d}(t-g+1), \ldots, \dot{\hat{\theta}}_\mathrm{d}(t-1), \dot{\hat{\theta}}_\mathrm{d}(t)]^\mathrm{T} \in \mathbb{R}^{L\times g}$, and the output weight $W_\mathrm{out}(t+1)$ is calculated according to (10.13).

In summary, in the above control scheme, the joint angle as input is fed into the cerebellum model, and the corresponding joint velocities are obtained after a cerebellum-like simulation time. The adopted ZND combined with RCM constraints is used to get the training signal to iteratively update the parameters of the PFs layer of the cerebellum model.

10.3.3 Visual Servoing Control Scheme

For the visual servoing task, both the kinematic model of the flexible endoscope and the image Jacobian matrix are difficult to model accurately. As with the tracking control task, this chapter introduces ZND for estimating the Jacobian matrix J_v in (10.6). A vector-valued error function is defined to measure the error:

$$\mathbf{e}_\mathrm{v}(t) = \dot{\mathbf{p}}_\mathrm{v}(t) - \hat{J}_\mathrm{v}(t)\dot{\vartheta}_\mathrm{d}(t) \in \mathbb{R}^2, \quad (10.23)$$

where the estimated Jacobian matrix $\hat{J}_{\rm v}(t)$ is used in

$$\dot{\vartheta}_{\rm d}(t) = \hat{J}_{\rm v}^{\dagger}(t)\dot{\varepsilon}_{\rm v}(t). \tag{10.24}$$

According to ZND design idea, an equation is defined as follows:

$$\dot{\mathbf{e}}_{\rm v}(t) = -\varsigma \mathbf{e}_{\rm v}(t), \tag{10.25}$$

where $\varsigma \in \mathbb{R}^{+}$ is for convergence adjustment and $\dot{\mathbf{e}}_{\rm v}(t)$ is the derivative of $\mathbf{e}_{\rm v}(t)$ with respect to time. Substituting (10.23) into (10.25) yields

$$\dot{\hat{J}}_{\rm v}(t) = (\ddot{\mathbf{p}}_{\rm v}(t) - \hat{J}_{\rm v}(t)\ddot{\vartheta}_{\rm d}(t) + \varsigma(\dot{\mathbf{p}}_{\rm v}(t) - \hat{J}_{\rm v}(t)\dot{\vartheta}_{\rm d}(t)))\dot{\vartheta}_{\rm d}^{\dagger}(t), \tag{10.26}$$

where $\dot{\hat{J}}_{\rm v}(t)$ represents the derivative of $\hat{J}_{\rm v}(t)$ with respect to t, $\ddot{\mathbf{p}}_{\rm v}(t)$ is the derivative of $\dot{\mathbf{p}}_{\rm v}(t)$, $\ddot{\vartheta}(t)$ is the derivative of $\dot{\vartheta}(t)$, and $\dot{\vartheta}^{\dagger}(t)$ represents the pseudo-inverse of the $\dot{\vartheta}(t)$.

The actuation input of the endoscope $\vartheta(t)$ is fed to the precerebellar nucleus, and like the tracking control scheme, $\dot{\vartheta}_{\rm d}(t)$ as the CFs is used to train the corresponding $W_{\rm out}$ of the PFs in the cerebellum model. The PCs output the control signal $\dot{\vartheta}_{\rm out}(t)$ to control the endoscope to complete the visual servoing task.

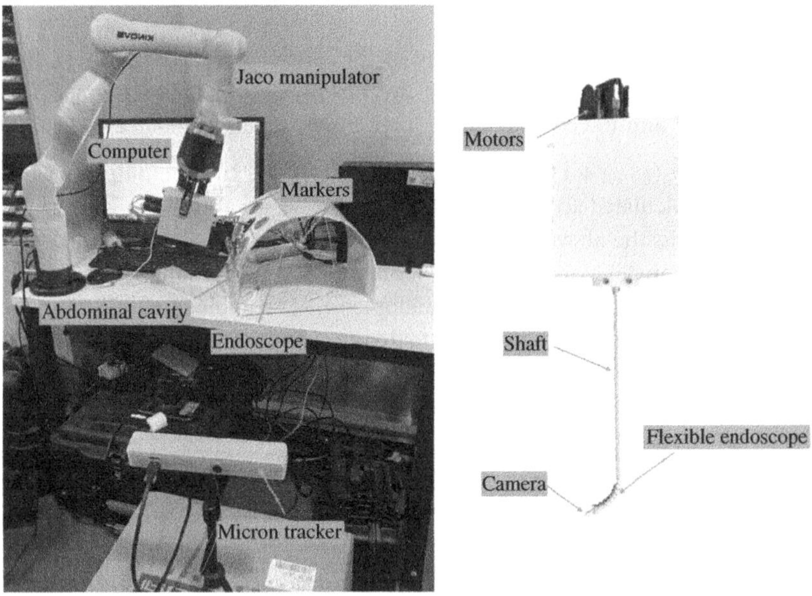

(a) (b)

Figure 10.4 Physical experiment platform with phantom abdominal cavity: (a) rigid-flexible hybrid robot; (b) flexible endoscope.

10.4 Verification

Physical experiment based on the Kinova Jaco Gen3 manipulator attached with a flexible endoscope is conducted to verify the efficacy of the control scheme as shown in Figure 10.4a. Figure 10.4b shows a flexible endoscope controlled by four motors. Two parameters are used to control two sets of motors, where one set of motors control the motion of the endoscope by pulling and another set by pushing.

10.4.1 Visual Servoing Task

The initial input of the flexible endoscope is set as $\vartheta(0) = [0, 0]^{\mathrm{T}}$ rad. The ZND design parameter of (10.26) is set as $\varsigma = 0.0003$. Experimental results of a visual servoing task are shown in Figure 10.5. From Figure 10.5a to 10.5b, it shows that although there is still some pixel error at the end, visually the visual servoing task is completed. This phenomenon is essentially due to the limitation of the hardware. That is, the accuracy of the movement is actually limited by the actuation resolution of the endoscope where the four motors pull the end flexible structure

(a) (b)

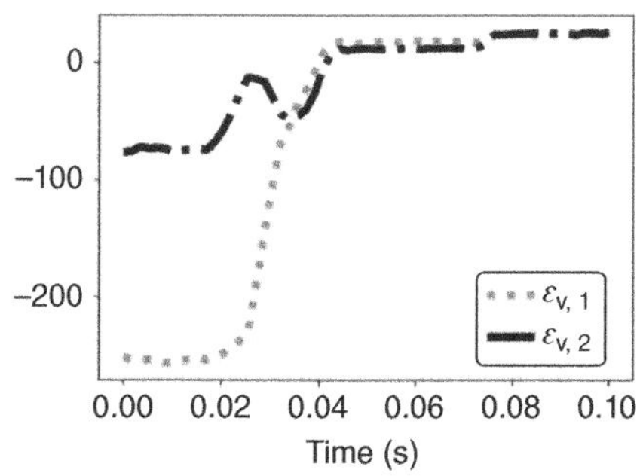

(c)

Figure 10.5 Experiment results of visual servoing of flexible endoscope: (a) initial image; (b) final image; (c) pixel errors.

to bend. The fabrication precision of the endoscope is fairly low, and the bending resolution is relatively big. This resulted in the fact that the pixel error in the endoscope visual servo experiment did not converge to zero. However, as shown in Figure 10.5c, the final pixel errors can meet the general surgical requirements.

10.5 Chapter Summary

This chapter presents a cerebellum-inspired network based on LSM. Then, based on this network, a model-free control scheme has been designed for rigid-flexible hybrid robots in MIS. The scheme is divided into two parts. The tracking control scheme is capable of dealing with two vital problems simultaneously, namely trajectory tracking and RCM constraints. The visual servoing control scheme controls the flexible endoscope for visual servoing tasks without the kinematics model. Experiments have been designed to verify the efficacy of the cerebellum-inspired scheme in the visual servoing task. The visual servoing task of the flexible endoscope can be accomplished successfully.

Part VII

Neurodynamics-based Model-free Teleoperation Method

11

Neurodynamics-based Teleoperation

11.1 Introduction

With a human operator in the control loop, teleoperation technology plays a pivotal role in human–robot interaction. Usually, the human operator is responsible for decision-making or high-level planning, while the robot handles the mechanical implementation. Teleoperation technology provides a promising way for human operators to perform remote tasks, offering significant practical value. For instance, teleoperation can protect humans from exposure to hazardous environments during missions in extreme environments. Moreover, teleoperation technology can also be applied to the medical realm, such as telesurgery, which may help alleviate the shortage of medical staff in less developed areas. Given its broad application prospects, devising a reliable, efficient, and portable teleoperation scheme emerges as an essential research question.

Particular attention has been given to the control of continuum robots with unknown kinematics. Despite the promising results achieved in model-free control, little research on the teleoperation of continuum robots with unknown kinematics has been reported. Without a human operator in the control loop, these model-free control methods can only ensure that the robot operates as preset. In comparison, teleoperation technology can provide more flexibility to robots and enable human–robot interaction.

This chapter focuses on the teleoperation of continuum robots. Significant advances in teleoperation technology for continuum robots have been achieved over the past decades. For example, Xie et al. [163] proposed a new haptic joystick to achieve the teleoperation of continuum-mechanism-based medical robots. With the growing interest in continuum robots, teleoperation of these robots has gained extensive attention [164]. However, most existing teleoperation schemes for continuum robots have two major disadvantages. First, existing schemes require the availability of the slave robot's kinematic model. However, modeling a robot, especially a continuum robot, in the three-dimensional task space can be highly complex and time-consuming, reducing the real-time performance of teleoperation. Second, the unpredictable deformation and wear of the slave robot will introduce kinematic uncertainties to the established model, adversely affecting the accuracy of teleoperation. This chapter aims at the development of a teleoperation scheme that is independent of the kinematic model of the slave robot to address the aforementioned problems.

The design of control system is a fundamental aspect of teleoperation systems. As a specialized recurrent neural network designed for time-dependent problems, zeroing neurodynamics (ZND) has been recognized as an effective tool for the kinematic control of continuum robots. Compared to other methods, ZND performs better in terms of convergence [165], noise tolerance [66], and hardware implementation [166]. This chapter presents a control method based on damping ZND (DampZND) to solve the time-dependent differential inverse kinematics of slave robots with unknown parameters and to achieve the adaptive teleoperation of continuum robots, combined with teleoperation technology. Finally, physical experiments using a cable-driven continuum robot are conducted to verify the teleoperation scheme.

11.2 Preliminary

Typically, a teleoperation system comprises three components [167]: a local site, where a human operator drives the master device; a remote workbench, where an integrated system with the slave robot interacts with the object entity; and a communication channel that connects the two sites. The overall framework of the teleoperation system is shown in Figure 11.1. In this chapter, a haptic device, the Geomagic Touch, is selected as the master device. The human operator holds the handle stylus to produce the commands by handle motion at the local site to control the slave robot, following the desired path. The communication channels

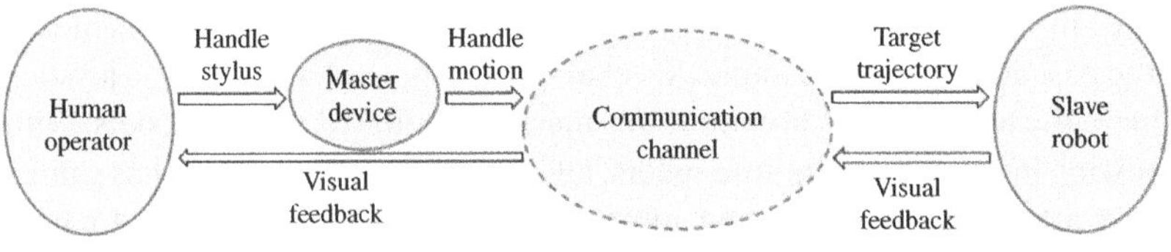

Figure 11.1 Overview of typical teleoperation system.

play a vital role in ensuring that the transmission of signals between the master–slave devices is effective, with time delay impacting the system's stability and accuracy [168]. Thus, the teleoperation system must be capable of coping with rapidly changing situations in real time. This chapter focuses on achieving adaptive teleoperation of continuum robots, assuming reliable communication channels and negligible time delay. Finally, the control system drives the slave robot to follow the desired path. Simultaneously, visual feedback on the robot's state is transmitted back to the human operator in real time, enabling appropriate decision-making.

11.3 Method

This section elaborates the theoretical foundation for the teleoperation of continuum robots, including the overview of a teleoperation system and a DampZND-based control system [169].

11.3.1 System Overview

To begin with, the schematic diagram of the teleoperation system is illustrated in Figure 11.2. Based on the real-time visual feedback information provided by the communication interface, the human operator decides how to manipulate the master device. When the handle stylus of the master device is moved, the motion trajectory of the handle, which serves as the desired path, is transmitted to the remote system through the communication interface. The communication interface can be implemented with one computer or two computers connected via the internet, depending on the specific requirements. The remote DampZND-based control system is crucial for the teleoperation system. It takes the desired path and

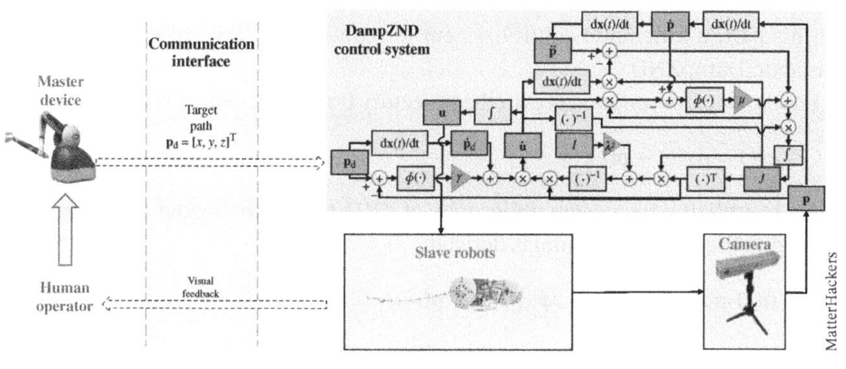

Figure 11.2 Schematic diagram of teleoperation system.

the actual trajectory of the slave robot as inputs and generates the actuation signal to control the slave robot. By fully utilizing the feedback information, the control system achieves adaptive control of the slave robot in the absence of a kinematic model. The detailed design process of the control system will be presented in the following section. Regarding the slave robot, continuum robots are considered. In the remote system, a MicronTracker H3-60 camera is used to measure the actual end-effector position of the slave robot, providing feedback information to the control system. Simultaneously, the real-time working scene of the slave robot is sent back to the human operator through the communication interface.

11.3.2 Robot Kinematics

Typically, the forward kinematics of a continuum robot is given by

$$\mathbf{p}(t) = \mathbf{f}(\mathbf{u}(t)), \tag{11.1}$$

where $\mathbf{p}(t) \in \mathbb{R}^m$ denotes the position of robot end-effector in the m-dimensional task space, $\mathbf{u}(t) \in \mathbb{R}^n$ denotes the state of n actuators of the continuum robot, and $\mathbf{f}(\cdot)$ denotes a nonlinear kinematic mapping function. Accordingly, the differential kinematics of the continuum robot is as follows:

$$\dot{\mathbf{p}}(t) = J(t)\dot{\mathbf{u}}(t), \tag{11.2}$$

where $\dot{\mathbf{p}}(t)$ denotes the first time derivative of $\mathbf{p}(t)$ and

$$J(t) = \frac{\partial \mathbf{f}(\mathbf{u}(t))}{\partial \mathbf{u}(t)} \in \mathbb{R}^{m\times n} \tag{11.3}$$

is the Jacobian matrix of the continuum robot.

11.3.3 DampZND-based Control System

In this part, a kinematic control system for continuum slave robots is designed based on DampZND.

Firstly, an error function $\mathbf{e}(t)$ in the trajectory tracking problem is constructed as

$$\mathbf{e}(t) = \mathbf{p}_{\mathrm{d}}(t) - \mathbf{p}(t). \tag{11.4}$$

Next, by utilizing the design philosophy of ZND, the following dynamic equation related to the actuation signal is derived:

$$\dot{\mathbf{u}}(t) = J^{\dagger}(t)\left(\dot{\mathbf{p}}_{\mathrm{d}}(t) + \gamma\boldsymbol{\phi}\left(\mathbf{p}_{\mathrm{d}}(t) - \mathbf{p}(t)\right)\right), \tag{11.5}$$

where $J^{\dagger}(t)$ is the pseudo-inverse of Jacobian matrix $J(t)$, $\dot{\mathbf{p}}_{\mathrm{d}}(t)$ is the desired velocity, γ is the convergence parameter of the ZND model, and $\boldsymbol{\phi}$ is an array of activation functions.

Solving the dynamic equation (11.5), one can obtain the actuation signal $\mathbf{u}(t)$. However, as mentioned earlier, model uncertainties may exist for any robot as a result of many factors, which will lead to the inaccuracy of $J(t)$, thereby degrading the performance of the control system. On the other hand, due to the structural and material characteristics of continuum robots, it is difficult to establish kinematic models for continuum robots, which will even lead to unknown Jacobian matrix $J(t)$.

To overcome the above problems, instead of analytically calculating the Jacobian matrix based on the kinematic model, this chapter attempts to utilize a ZND model to estimate the Jacobian matrix. Therefore, another error function is defined as follows:

$$\zeta(t) = \dot{\mathbf{p}}(t) - J(t)\dot{\mathbf{u}}(t), \tag{11.6}$$

where $\dot{\mathbf{p}}(t)$ denotes the actual velocity of the end-effector and $J(t)\dot{\mathbf{u}}(t)$ can be deemed as the estimated end-effector velocity. Then, utilizing the design philosophy of ZND again, the dynamic equation for Jacobian estimation is obtained as follows:

$$\dot{J}(t) = (\ddot{\mathbf{p}}(t) - J(t)\ddot{\mathbf{u}}(t) + \mu\boldsymbol{\phi}(\dot{\mathbf{p}}(t) - J(t)\dot{\mathbf{u}}(t)))\,\dot{\mathbf{u}}^{\dagger}(t), \tag{11.7}$$

where μ denotes another convergence parameter.

Generally, the tracking control problem of a slave robot with unknown kinematics can be solved by the ZND models (11.5) and (11.7). It is intuitively observed that both the models require the pseudo-inverse operation. However, the pseudo-inverse operation tends to have stability problems in the neighborhoods of singularities. Under the circumstance the resulting configuration of robot is close to the singularity, the pseudo-inverse method will lead to enormous locomotion in actuation space, where the robot motion is not well-behaved. Hence, the solution strategy of the pseudo-inverse calculation is very important to the stability of the teleoperation system.

This chapter presents DampZND models that can effectively avoid the problems of the pseudo-inverse method with singularities and provides a numerically stable method of selecting actuation signal [169]. The ZND model (11.5) is improved to the following DampZND model:

$$\dot{\mathbf{u}}(t) = J_{\mathrm{dp}}^{-1}(t)\,(\dot{\mathbf{p}}_{\mathrm{d}}(t) + \gamma\boldsymbol{\phi}(\mathbf{p}_{\mathrm{d}}(t) - \mathbf{p}(t))), \tag{11.8}$$

where J_{dp}^{-1} is the inverse solution including a damping factor λ, whose specific formula is

$$J_{\mathrm{dp}}^{-1} = J^{\mathrm{T}}\left(JJ^{\mathrm{T}} + \lambda^2 I\right)^{-1} \tag{11.9}$$

with I being the identity matrix. Similarly, the ZND model (11.7) is also improved to the following DampZND model:

$$\dot{J}(t) = \left(\ddot{\mathbf{p}}(t) - J(t)\ddot{\mathbf{u}}(t) + \mu\boldsymbol{\phi}(\dot{\mathbf{p}}(t) - J(t)\dot{\mathbf{u}}(t))\right)\dot{\mathbf{u}}_{\mathrm{dp}}^{-1}(t), \quad (11.10)$$

where $\dot{\mathbf{u}}_{\mathrm{dp}}^{-1}$ is also the inverse solution including a damping factor λ, which satisfies

$$\dot{\mathbf{u}}_{\mathrm{dp}}^{-1} = (\dot{\mathbf{u}}^{\mathrm{T}}\dot{\mathbf{u}} + \lambda^2 I)^{-1}\dot{\mathbf{u}}^{\mathrm{T}}. \quad (11.11)$$

Finally, the control system for continuum slave robots is established based on DampZND models (11.8) and (11.10), and the diagram can be found in Figure 11.2. The DampZND control system is applicable to the continuum robot with unknown kinematics. Accordingly, the adaptive teleoperation of continuum robots is achieved with the aid of the DampZND control system.

11.3.4 Theoretical Analysis

The singular value decomposition (SVD) provides a powerful method for stability analysis of the DampZND approach in this section.

A SVD of the Jacobian matrix J is expressed in the form:

$$J = ADB^{\mathrm{T}} \quad (11.12)$$

where $A \in \mathbb{R}^{m\times m}$ and $B \in \mathbb{R}^{n\times n}$ are orthogonal matrices, and $D \in \mathbb{R}^{m\times n}$ is diagonal. The only nonzero elements of matrix D are the value $\beta_i = d_{i,i}$ along the diagonal. Without loss of generality, suppose $m \leq n$ and

$$\beta_1 \geq \beta_2 \geq \cdots \geq \beta_m \geq 0. \quad (11.13)$$

It is worth noting that the value β_i can be zero. Thus, the rank of matrix J is equal to the largest number k such that $\beta_k \neq 0$ in matrix D, and $\beta_i = 0$ when $i > k$. Besides, we use $\mathbf{a}_i$ and $\mathbf{b}_i$ to represent the ith column vector of A and B. The orthogonality of A and B suggests that the corresponding columns of A (resp., of B) constitute an orthogonal basis for $\mathbb{R}^m$ (resp., for $\mathbb{R}^n$). The column vectors $\mathbf{b}_{k+1}, \dots, \mathbf{b}_n$ are an orthogonal basis for the null space of matrix J.

Obviously, the SVD of J always exists, which means that J can be constructed as follows:

$$\begin{aligned} J &= \sum_{i=1}^{m} \beta_i \mathbf{a}_i \mathbf{b}_i^{\mathrm{T}} \\ &= \sum_{i=1}^{k} \beta_i \mathbf{a}_i \mathbf{b}_i^{\mathrm{T}}. \end{aligned} \quad (11.14)$$

The transpose matrix D^{T} of D is the dimension of $n \times m$ with diagonal entries $\beta_i = d_{i,i}$. The product $D^{\mathrm{T}}D$ is an $m \times m$ matrix with the diagonal value $d_{i,i}^2$. As a result, the DampZND method can be intuitively understood with SVD. The final matrix $J^{\mathrm{T}}J + \lambda^2 I$ is expressed in

$$\begin{aligned} JJ^{\mathrm{T}} + \lambda^2 I &= \left(ADB^{\mathrm{T}}\right)\left(BD^{\mathrm{T}}A^{\mathrm{T}}\right) + \lambda^2 I \\ &= ADD^{\mathrm{T}}A^{\mathrm{T}} + \lambda^2 \\ &= A\left(DD^{\mathrm{T}} + \lambda^2 I\right)A^{\mathrm{T}}, \end{aligned} \tag{11.15}$$

where

$$B^{\mathrm{T}}B = I \tag{11.16}$$

due to the properties of orthogonal matrix.

The matrix $DD^{\mathrm{T}} + \lambda^2 I$ is also a diagonal matrix with specific entries $\beta_i^2 + \lambda^2$. Surprisingly, the item $DD^{\mathrm{T}} + \lambda^2 I$ is a non-singular matrix, whose inverse is the $m \times m$ diagonal form with nonzero values $(\beta_i^2 + \lambda^2)^{-1}$. Therefore, the following formula is obtained:

$$\begin{aligned} J^{\mathrm{T}}(JJ^{\mathrm{T}} + \lambda^2 I)^{-1} &= BD^{\mathrm{T}}(DD^{\mathrm{T}} + \lambda^2 I)^{-1}A^{\mathrm{T}} \\ &= BEA^{\mathrm{T}}, \end{aligned} \tag{11.17}$$

where E is the $n \times m$ diagonal matrix with each element being

$$x_{i,i} = \frac{\beta_i}{\beta_i^2 + \lambda^2}. \tag{11.18}$$

Thus, the DampZND solution is rewritten as follows:

$$J^{\mathrm{T}}(JJ^{\mathrm{T}} + \lambda^2 I)^{-1} = \sum_{i=1}^{\mathrm{T}} \frac{\beta_i}{\beta_i^2 + \lambda^2}\mathbf{b}_i\mathbf{a}_i^{\mathrm{T}}. \tag{11.19}$$

In contrast, the formula can be similarly derived based on the pseudo-inverse method as follows:

$$J^{\dagger} = \sum_{i=1}^{t} \beta_i^{-1}\mathbf{b}_i\mathbf{a}_i^{\mathrm{T}}. \tag{11.20}$$

In both cases above, the matrix J is inverted by a unified mathematical expression $\sum_i \mu_i \mathbf{b}_i\mathbf{a}_i^{\mathrm{T}}$. For the pseudo-inverse, the value of μ_i is just β_i^{-1}, whereas for DampZND method, $\mu_i = \beta_i/(\beta_i^2 + \lambda^2)$. In fact, the pseudo-inverse is not stable when β_i tends to zero, which in this sense makes the robot motion appear as great oscillation in the teleoperation system.

Above all, the DampZND can effectively smooth out the performance of pseudo-inverse method in the neighborhood of singularities.

11.4 Verification

In this section, the teleoperation experiment of a continuum robot is performed to verify that the system can achieve the adaptive teleoperation of continuum robots.

11.4.1 Setting

The experiment platform is depicted in Figure 11.3. A marker is used to assist the camera to measure the position of the end-effector of the robot. The task is to teleoperate the continuum robot to move along a preset trajectory. We use the computer as the communication interface, and the DampZND control system is implemented via Simulink. The parameters of the control system are set as $\gamma = 6$, $\mu = 0.1$, and $\lambda = 0.1$.

11.4.2 Results

Some snapshots of the teleoperation of the continuum robot are displayed in Figure 11.4. The comparison between the handle trajectory and the robot trajectory is illustrated in Figure 11.5. It can be seen that the teleoperation of the continuum robot is fulfilled with the system. The maximum tracking error in this task is less than 3×10^{-3} m, and the root-mean-square error (RMSE) is 1.031×10^{-3} m. In summary, the above results have revealed the efficacy of the adaptive teleoperation system for the continuum robot.

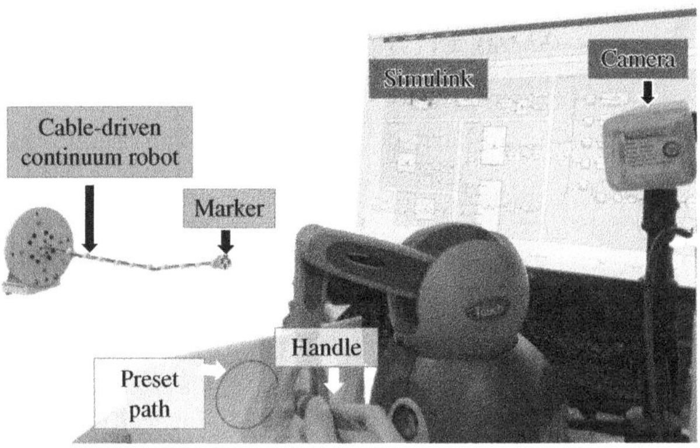

Figure 11.3 Overview of physical platform for teleoperation of cable-driven continuum robot.

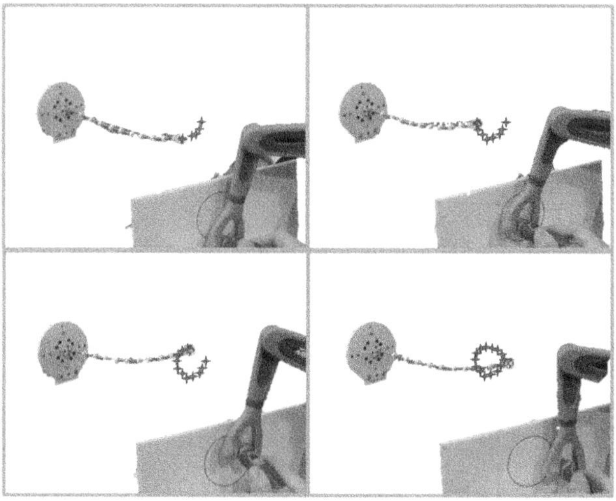

Figure 11.4 Some snapshots of teleoperation of continuum robot.

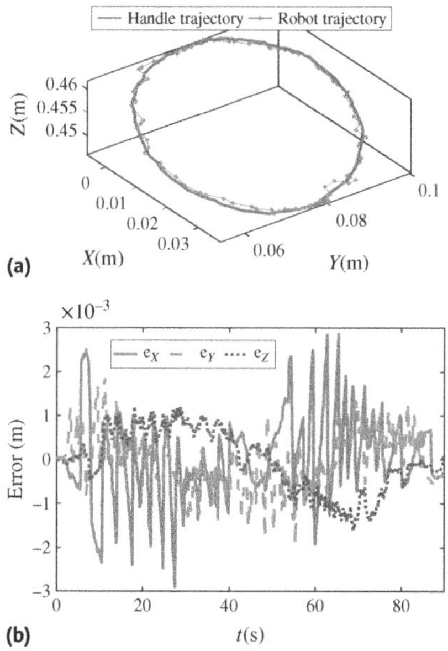

Figure 11.5 Teleoperation results of continuum robot: (a) trajectory comparison; (b) position errors.

11.5 Chapter Summary

This chapter presents the design of a DampZND control system and a DampZND-based teleoperation system for human–robot interaction. With the aid of the teleoperation system, human operators can control the continuum robot without prior knowledge of the kinematic model. Additionally, experiments conducted on a cable-driven continuum robot have validated the efficacy of the adaptive teleoperation system. The results have illustrated that the continuum robot can follow the path specified by users accurately.

References

1 Jianhua Li, Dingjia Li, Chongyang Wang, Wei Guo, Zhidong Wang, Zhongtao Zhang, and Hao Liu. Active collision avoidance for teleoperated multi-segment continuum robots toward minimally invasive surgery. *The International Journal of Robotics Research*, 43(7): 918–941, 2024.

2 Xin Dong, David Palmer, Dragos Axinte, and James Kell. In-situ repair/maintenance with a continuum robotic machine tool in confined space. *Journal of Manufacturing Processes*, 38: 313–318, 2019.

3 Azamat Yeshmukhametov, Koichi Koganezawa, Yoshio Yamamoto, Zholdas Buribayev, Zhassuzak Mukhtar, and Yedilkhan Amirgaliyev. Development of continuum robot arm and gripper for harvesting cherry tomatoes. *Applied Sciences-Basel*, 12(14): 6922, 2022.

4 Xin Dong, Mingfeng Wang, Abdelkhalick Mohammad, Weiming Ba, Matteo Russo, Andy Norton, James Kell, and Dragos Axinte. Continuum robots collaborate for safe manipulation of high-temperature flame to enable repairs in challenging environments. *IEEE/ASME Transactions on Mechatronics*, 27(5): 4217–4220, 2022.

5 Bin He, Zhipeng Wang, Qiang Li, Hong Xie, and Runjie Shen. An analytic method for the kinematics and dynamics of a multiple-backbone continuum robot. *International Journal of Advanced Robotic Systems*, 10(1): 84, 2013.

6 Xinjia Huang, Jiang Zou, and Guoying Gu. Kinematic modeling and control of variable curvature soft continuum robots. *IEEE/ASME Transactions on Mechatronics*, 26(6): 3175–3185, 2021.

7 Priyanka Rao, Chloe Pogue, Quentin Peyron, Eric Diller, and Jessica Burgner-Kahrs. Modeling and analysis of tendon-driven continuum robots for rod-based locking. *IEEE Robotics and Automation Letters*, 8(6): 3126–3133, 2023.

8 Dong Wang, Chengru Jiang, and Guoying Gu. Modeling and design of lattice-reinforced pneumatic soft robots. *IEEE Transactions on Robotics*, 40: 606–623, 2024.

9 Cosimo Della Santina, Robert Kevin Katzschmann, Antonio Bicchi, and Daniela Rus. Model-based dynamic feedback control of a planar soft robot: Trajectory tracking and interaction with the environment. *The International Journal of Robotics Research*, 39(4): 490–513, 2020.

10 Olivier Goury and Christian Duriez. Fast, generic, and reliable control and simulation of soft robots using model order reduction. *IEEE Transactions on Robotics*, 34(6): 1565–1576, 2018.

11 Xinran Wang and Nicolas Rojas. Cosserat rod modeling and validation for a soft continuum robot with self-controllable variable curvature. In *IEEE 7th International Conference on Soft Robotics*, IEEE, pages 705–710, 2024.

12 Zheng Chen, Hui Ren, Wei Fan, and Ping Zhou. Closed-form solutions for the planar hard-magnetic soft beam with large deformations and its application to soft continuum robots. *Nonlinear Dynamics*, 113(7): 6157–6180, 2025.

13 Junius Santoso and Cagdas Denizel Onal. An origami continuum robot capable of precise motion through torsionally stiff body and smooth inverse kinematics. *Soft Robotics*, 8(4): 371–386, 2021. https://doi.org/10.1089/soro.2020.0026.

14 Leopoldo Angrisani, Stanislao Grazioso, Giuseppe Di Gironimo, Dario Panariello, and Annarita Tedesco. On the use of soft continuum robots for remote measurement tasks in constrained environments: A brief overview of applications. In *IEEE International Symposium on Measurements & Networking*, IEEE, pages 1–5, 2019.

15 Zixi Chen, Alexia Le Gall, Lorenzo Mocellin, Matteo Bernabei, Théo Dangel, Gastone Ciuti, Matteo Cianchetti, and Cesare Stefanini. Data models applied to soft robot modeling and control: A review. In *IEEE Transactions on Automation Science and Engineering*, 22: 2241–2256, 2025. https://doi.org/10.1109/TASE.2024.3377291.

16 Delin Hu, Francesco Giorgio-Serchi, Shiming Zhang, and Yunjie Yang. Stretchable e-skin and transformer enable high-resolution morphological reconstruction for soft robots. *Nature Machine Intelligence*, 5(3): 261–272, 2023. ISSN 2522-5839. https://doi.org/10.1038/s42256-023-00622-8.

17 Kathrin Nuelle, Tim Sterneck, Sven Lilge, Dezhu Xiong, Jessica Burgner-Kahrs, and Tobias Ortmaier. Modeling, calibration, and evaluation of a tendon-actuated planar parallel continuum robot. *IEEE Robotics and Automation Letters*, 5(4): 5811–5818, 2020.

18 Peng Yu, Ning Tan, Yuyang Wu, Binbin Qiu, and Kai Huang. Model-free synchronous motion generation of multiple heterogeneous continuum robots. *IEEE Transactions on Industrial Informatics*, 20(3): 3209–3221, 2024. ISSN 19410050.

19 Thomas George Thuruthel, Egidio Falotico, Mariangela Manti, Andrea Pratesi, Matteo Cianchetti, and Cecilia Laschi. Learning closed loop kinematic controllers for continuum manipulators in unstructured environments. *Soft Robotics*, 4(3): 285–296, 2017.

20 Mohammadreza Kasaei, Keyhan Kouhkiloui Babarahmati, Zhibin Li, and Mohsen Khadem. Data-efficient non-parametric modelling and control of an extensible soft manipulator. In *International Conference on Robotics and Automation*, pages 2641–2647, 2023.

21 Yunong Zhang, Lin Xiao, Zhengli Xiao, and Mingzhi Mao. *Zeroing Dynamics, Gradient Dynamics, and Newton Iterations*. Boca Raton, FL, USA: CRC Press, 2016.

22 Zhijun Zhang, Lunan Zheng, Tairu Qiu, and Feiqi Deng. Varying-parameter convergent-differential neural solution to time-varying overdetermined system of linear equations. *IEEE Transactions on Automatic Control*, 65(2): 874–881, 2020.

23 Lunan Zheng and Zhijun Zhang. Time-varying quadratic-programming-based error redefinition neural network control and its application to mobile redundant manipulators. *IEEE Transactions on Automatic Control*, 67(11): 6151–6158, 2022.

24 Jie Jin, Weijie Chen, Chaoyang Chen, and Lianghong Wu. Fixed-time solution of inequality constrained time-varying linear systems via zeroing neural networks. *Journal of Franklin Institute*, 361(9): 106870, 2024.

25 Lin Xiao, Yunrui Xie, Qiuyue Zuo, Ping Tan, Ping Liu, and Yongjun He. Design and analysis of finite-time convergent complex-valued zeroing neural networks with application to time-variant complex matrix inversion. *Information Sciences*, 681: 121217, 2024.

26 Robert James Webster and Bryan Allen Jones. Design and kinematic modeling of constant curvature continuum robots: A review. *The International Journal of Robotics Research*, 29(13): 1661–1683, 2010.

27 Yunong Zhang and Dongsheng Guo. *Zhang Functions and Various Models*. Berlin, Heidelberg: Springer, 2015. ISBN 9783662473344.

28 Chenfu Yi, Yuhuan Chen, and Xinhua Lan. Comparison on neural solvers for the Lyapunov matrix equation with stationary & nonstationary coefficients. *Applied Mathematical Modelling*, 37(4): 2495–2502, 2013.

29 Yunong Zhang, Ke Chen, Xuezhong Li, Chengfu Yi, and Hong Zhu. Simulink modeling and comparison of Zhang neural networks and gradient neural networks for time-varying Lyapunov equation solving. In *International Conference on Natural Computation*, pages 521–525, 2008.

30 Albrecht Böttcher and David Wenzel. The Frobenius norm and the commutator. *Linear Algebra and its Applications*, 429(8-9): 1864–1885, 2008.

31 Taoming Liu, Russell Jackson, Dominique Franson, Nate Lombard Poirot, Reinhardt Kam Criss, Nicole Seiberlich, Mark A Griswold, and M Cenk Çavuşoğlu. Iterative Jacobian-based inverse kinematics and open-loop control of an MRI-guided magnetically actuated steerable catheter system. *IEEE/ASME Transactions on Mechatronics*, 22(4): 1765–1776, 2017.

32 Salvador Cobos-Guzman, David Palmer, and Dragos Axinte. Kinematic model to control the end-effector of a continuum robot for multi-axis processing. *Robotica*, 35(1): 224–240, 2017.

33 Quang-Cuong Pham. A general, fast, and robust implementation of the time-optimal path parameterization algorithm. *IEEE Transactions on Robotics*, 30(6): 1533–1540, 2014.

34 Mingxing Yuan, Zheng Chen, Bin Yao, and Xiaocong Zhu. Time optimal contouring control of industrial biaxial gantry: A highly efficient analytical solution of trajectory planning. *IEEE/ASME Transactions on Mechatronics*, 22(1): 247–257, 2016.

35 Cosimo Della Santina, Antonio Bicchi, and Daniela Rus. On an improved state parametrization for soft robots with piecewise constant curvature and its use in model based control. *IEEE Robotics and Automation Letters*, 5(2): 1001–1008, 2020. ISSN 23773766.

36 Achille Melingui, Othman Lakhal, Boubaker Daachi, Jean Bosco Mbede, and Rochdi Merzouki. Adaptive neural network control of a compact bionic handling arm. *IEEE/ASME Transactions on Mechatronics*, 20(6): 2862–2875, 2015.

37 Wen-Shyong Yu and Chien-Chih Weng. An observer-based adaptive neural network tracking control of robotic systems. *Applied Soft Computing*, 13(12): 4645–4658, 2013.

38 Yunong Zhang, Yiwen Yang, Ning Tan, and Binghuang Cai. Zhang neural network solving for time-varying full-rank matrix Moore–Penrose inverse. *Computing*, 92(2): 97–121, 2011.

39 Dongsheng Guo and Yunong Zhang. Li-function activated ZNN with finite-time convergence applied to redundant-manipulator kinematic control via time-varying Jacobian matrix pseudoinversion. *Applied Soft Computing*, 24: 158–168, 2014.

40 Long Jin, Yunong Zhang, Shuai Li, and Yinyan Zhang. Modified ZNN for time-varying quadratic programming with inherent tolerance to noises and its application to kinematic redundancy resolution of robot manipulators. *IEEE Transactions on Industrial Electronics*, 63(11): 6978–6988, 2016.

41 Ning Tan, Mingwei Huang, Peng Yu, and Tao Wang. Neural-dynamics-enabled Jacobian inversion for model-based kinematic control of multi-section continuum manipulators. *Applied Soft Computing*, 103: 107114, 2021. ISSN 15684946.

42 Zhiguo Tan, Yueming Hu, Lin Xiao, and Ke Chen. Robustness analysis and robotic application of combined function activated RNN for time-varying matrix pseudo inversion. *IEEE Access*, 7: 33434–33440, 2019.

43 Yu-Nong Zhang, Zhen Li, Dong-Sheng Guo, Ke Chen, and Pei Chen. Superior robustness of using power-sigmoid activation functions in Z-type models for time-varying problems solving. In *International Conference on Machine Learning and Cybernetics, volume 2,* IEEE, pages 759–764, 2013.

44 Shuai Li, Sanfeng Chen, and Bo Liu. Accelerating a recurrent neural network to finite-time convergence for solving time-varying Sylvester equation by using a sign-bi-power activation function. *Neural Processing Letters*, 37(2): 189–205, 2013.

45 Lin Xiao, Lei Jia, Yongsheng Zhang, Zeshan Hu, and Jianhua Dai. Finite-time convergence and robustness analysis of two nonlinear activated ZNN models for time-varying linear matrix equations. *IEEE Access*, 7: 135133–135144, 2019.

46 Long Jin, Shuai Li, Huanqing Wang, and Zhijun Zhang. Nonconvex projection activated zeroing neurodynamic models for time-varying matrix pseudoinversion with accelerated finite-time convergence. *Applied Soft Computing*, 62: 840–850, 2018.

47 Zhiguo Tan, Weibing Li, Lin Xiao, and Yueming Hu. New varying-parameter ZNN models with finite-time convergence and noise suppression for time-varying matrix Moore–Penrose inversion. *IEEE Transactions on Neural Networks and Learning Systems*, 31(8): 2980–2992, 2020.

48 Sven Lilge and Jessica Burgner-Kahrs. Kinetostatic modeling of tendon-driven parallel continuum robots. *IEEE Transactions on Robotics*, 39(2): 1563–1579, 2023. ISSN 19410468. https://doi.org/10.1109/TRO.2022.3226157.

49 Ryota Morimoto, Satoshi Nishikawa, Ryuma Niiyama, and Yasuo Kuniyoshi. Model-free reinforcement learning with ensemble for a soft continuum robot arm. In *IEEE 4th International Conference on Soft Robotics,* IEEE, pages 141–148, 2021.

50 Yusong Jin, Yufei Wang, Xiaotong Chen, Zhanchi Wang, Xinghua Liu, Hao Jiang, and Xiaoping Chen. Model-less feedback control for soft manipulators. In *IEEE International Conference on Intelligent Robots and Systems,* pages 2916–2922, 2017.

51 Minhan Li, Rongjie Kang, David Branson, and Jian Sheng Dai. Model-free control for continuum robots based on an adaptive Kalman filter. *IEEE/ASME Transactions on Mechatronics*, 23(1): 286–297, 2018.

52 Pengcheng Liu, Hongnian Yu, and Shuang Cang. Adaptive neural network tracking control for underactuated systems with matched and mismatched disturbances. *Nonlinear Dynamics*, 98(2): 1447–1464, 2019.

53 Lin Xiao, Zhijun Zhang, Zili Zhang, Weibing Li, and Shuai Li. Design, verification and robotic application of a novel recurrent neural network for computing dynamic Sylvester equation. *Neural Networks*, 105: 185–196, 2018.

54 Zhijun Zhang and Lunan Zheng. A complex varying-parameter for solving online time-varying complex Sylvester equation. *IEEE Transactions on Cybernetics*, 49(10): 3627–3639, 2019.

55 Zhijun Zhang, Lunan Zheng, Jian Weng, Yijun Mao, Wei Lu, and Lin Xiao. A new varying-parameter recurrent neural-network for online solution of time-varying Sylvester equation. *IEEE Transactions on Cybernetics*, 48(11): 3135–3148, 2018.

56 Bolin Liao, Yunong Zhang, and Long Jin. Discretization of ZNN models for dynamic equality-constrained quadratic programming with application to manipulators. *IEEE Transactions on Neural Networks and Learning Systems*, 27(2): 225–237, 2016.

57 Yunong Zhang, Ke Chen, and Hong Zhou Tan. Performance analysis of gradient neural network exploited for online time-varying matrix inversion. *IEEE Transactions on Automatic Control*, 54(8): 1940–1945, 2009.

58 Chenfu Yi, Yuhuan Chen, and Zhongliang Lu. Improved gradient-based neural networks for online solution of Lyapunov matrix equation. *Information Processing Letters*, 111(16): 780–786, 2011.

59 Yunong Zhang, Yiwen Yang, and Gongqin Ruan. Performance analysis of gradient neural network exploited for online time-varying quadratic minimization and equality-constrained quadratic programming. *Neurocomputing*, 74(10): 1710–1719, 2011.

60 Shuai Li, Ruofan Kong, and Yi Guo. Cooperative distributed source seeking by multiple robots: Algorithms and experiments. *IEEE/ASME Transactions on Mechatronics*, 19(6): 1810–1820, 2014.

61 Long Jin and Shuai Li. Distributed task allocation of multiple robots: A control perspective. *IEEE Transactions on Systems, Man, and Cybernetics: Systems*, 48(5): 693–701, 2018.

62 Peter Corke. *Robotics, Vision and Control*. Springer, 2011.

63 Peng Yu, Ning Tan, and Zhiyan Zhong. Comparative studies and performance analysis on neural-dynamics-driven control of redundant robot manipulators with unknown models. *Engineering Applications of Artificial Intelligence*, 117: 105528, 2023. ISSN 0952-1976.

64 Robert Katzschmann, Cosimo Della Santina, Yasunori Toshimitsu, Antonio Bicchi, and Daniela Rus. Dynamic motion control of multi-segment soft robots using piecewise constant curvature matched with an augmented rigid body model. In *IEEE International Conference on Soft Robotics*, pages 454–461, 2019.

65 Zheyuan Gong, Xi Fang, Xingyu Chen, Jiahui Cheng, Zhexin Xie, Jiaqi Liu, Bohan Chen, Hui Yang, Shihan Kong, Yufei Hao, Tianmiao Wang, Junzhi Yu, and Li Wen. A soft manipulator for efficient delicate grasping in shallow water: Modeling, control, and real-world experiments. *The International Journal of Robotics Research*, 40(1): 449–469, 2021.

66 Long Jin, Yunong Zhang, Shuai Li, and Yinyan Zhang. Noise-tolerant ZNN models for solving time-varying zero-finding problems: A control-theoretic approach. *IEEE Transactions on Automatic Control*, 62(2): 992–997, 2017.

67 Zhigang Liu, Xin Luo, and Zidong Wang. Convergence analysis of single latent factor-dependent, nonnegative, and multiplicative update-based nonnegative latent factor models. *IEEE Transactions on Neural Networks and Learning Systems*, 32(4): 1737–1749, 2021.

68 Ning Tan, Peng Yu, Zhaohui Zhong, and Yunong Zhang. Data-driven control for continuum robots based on discrete zeroing neural networks. *IEEE Transactions on Industrial Informatics*, 19(5): 7088–7098, 2023. ISSN 19410050. https://doi.org/10.1109/TII.2022.3204307.

69 Peng Yu, Ning Tan, Mao Zhang, and Fenglei Ni. Inverse-free tracking control of continuum robots with unknown models based on gradient neural networks. In *IEEE 11th Data Driven Control and Learning Systems Conference*, pages 648–653, 2022. https://doi.org/10.1109/DDCLS55054.2022.9858510.

70 Yingzhong Tian, Mingxuan Luan, Xu Gao, Wenbin Wang, and Long Li. Kinematic analysis of continuum robot consisted of driven flexible rods. *Mathematical Problems in Engineering*, 2016(1): 6984194, 2016.

71 Jessica Burgner-Kahrs, Daniel Caleb Rucker, and Howie Choset. Continuum robots for medical applications: A survey. *IEEE Transactions on Robotics*, 31(6): 1261–1280, 2015.

72 Mei Liu, Jialiang Fan, Yu Zheng, Shuai Li, and Long Jin. A simultaneous learning and control scheme for redundant manipulators with physical constraints on decision variable and its derivative. *IEEE Transactions on Industrial Electronics*, 69(10): 10301–10310, 2022.

73 Kit-Hang Lee, Kin Chung Denny Fu, Martin Chun Wing Leong, Marco Chow, Hing-Choi Fu, Kaspar Althoefer, Kam Yim Sze, Chung-Kwong Yeung, and Ka-Wai Kwok. Nonparametric online learning control for soft continuum robot: An enabling technique for effective endoscopic navigation. *Soft Robotics*, 4(4): 324–337, 2017.

74 Haodong Wang, Wenyu Liang, Boyuan Liang, Hongliang Ren, Zhijiang Du, and Yan Wu. Robust position control of a continuum manipulator based on selective approach and Koopman operator. *IEEE Transactions on Industrial Electronics*, 70(12): 12522–12532, 2023.

75 Thomas George Thuruthel, Yasmin Ansari, Egidio Falotico, and Cecilia Laschi. Control strategies for soft robotic manipulators: A survey. *Soft Robotics*, 5(2): 149–163, 2018.

76 Barry Trimmer. Soft robot control systems: A new grand challenge? *Soft Robotics*, 1(4): 231–232, 2014.

77 Yimin Wei, Jianfeng Cai, and Michael Kwok-Po Ng. Computing Moore–Penrose inverses of Toeplitz matrices by Newton's iteration. *Mathematical and Computer Modelling*, 40(1-2): 181–191, 2004.

78 Yimin Wei. Successive matrix squaring algorithm for computing the Drazin inverse. *Applied Mathematics and Computation*, 108(2): 67–75, 2000. ISSN 0096-3003. https://doi.org/10.1016/S0096-3003(98)10118-2.

79 Jie Zhou, Yunmin Zhu, X Rong Li, and Zhisheng You. Variants of the Greville formula with applications to exact recursive least squares. *Siam Journal on Matrix Analysis & Applications*, 24(1): 150–164, 2002.

80 Yunong Zhang, Long Jin, Dongsheng Guo, Yonghua Yin, and Yao Chou. Taylor-type 1-step-ahead numerical differentiation rule for first-order derivative approximation and ZNN discretization. *Journal of Computational & Applied Mathematics*, 273: 29–40, 2015.

81 Shuai Li, Huanqing Wang, and Muhammad Usman Rafique. A novel recurrent neural network for manipulator control with improved noise tolerance. *IEEE Transactions on Neural Networks and Learning Systems*, 29(5): 1908–1918, 2017.

82 Mei Liu and Mingsheng Shang. Orientation tracking incorporated multicriteria control for redundant manipulators with dynamic neural network. *IEEE Transactions on Industrial Electronics*, 71(4): 3801–3810, 2023.

83 Zhengtai Xie, Long Jin, Xiujuan Du, Xiuchun Xiao, Hongxin Li, and Shuai Li. On generalized RMP scheme for redundant robot manipulators aided with dynamic neural networks and nonconvex bound constraints. *IEEE Transactions on Industrial Informatics*, 15(9): 5172–5181, 2019.

84 Zhijun Zhang, Lunan Zheng, Junming Yu, Yuanqing Li, and Zhuliang Yu. Three recurrent neural networks and three numerical methods for solving a repetitive motion planning scheme of redundant robot manipulators. *IEEE/ASME Transactions on Mechatronics*, 22(3): 1423–1434, 2017.

85 Nini Shi, Min Yang, Huanchang Huang, Ning Tan, and Yunong Zhang. Discrete-time ODE solutions generated by TZD and zead4ig2_y formulas: Numerical results. In *9th International Conference on Information Science and Technology*, IEEE, pages 304–310, 2019.

86 Muhammad Usama Goher, Peng Yu, Qinghua Lu, and Ning Tan. Acceleration-layer robotic control based on neural dynamics. In *IEEE Transactions on Industrial Electronics*, 72(11): 11434–11444, 2025. https://doi.org/10.1109/TIE.2025.3558038.

87 Loup Verlet. Computer "experiments" on classical fluids. I. Thermodynamical properties of Lennard-Jones molecules. *Physical Review*, 159(1): 98, 1967.

88 Zhuosong Fu, Yunong Zhang, and Weibing Li. Solving future nonlinear equation system via ZNN and novel general ILR3S formula with multitype manipulator applications. *IEEE Transactions on Industrial Electronics*, 71(10): 12623–12633, 2024. https://doi.org/10.1109/TIE.2024.3357867.

89 Ning Tan, Muhammad Usama Goher, Peng Yu, Zhenglong Sun, and Xiaoyi Gu. Discrete quad neural dynamics for inverse-free control of model-unavailable continuum robots. *IEEE Transactions on Industrial Electronics*, 71(1): 12601–12611, 2024. ISSN 15579948. https://doi.org/10.1109/TIE.2024.3357892.

90 Richard Seeber, Hernan Haimovich, and Martin Horn. Robust exact differentiators with predefined convergence time. *Automatica*, 134: 109858, 2021.

91 Bolin Liao and Yunong Zhang. From different ZFs to different ZNN models accelerated via Li activation functions to finite-time convergence for time-varying matrix pseudoinversion. *Neurocomputing*, 133: 512–522, 2014.

92 Jianrong Chen and Yunong Zhang. Online singular value decomposition of time-varying matrix via zeroing neural dynamics. *Neurocomputing*, 383: 314–323, 2020.

93 Dongsheng Guo, Zhuoyun Nie, and Laicheng Yan. Novel discrete-time Zhang neural network for time-varying matrix inversion. *IEEE Transactions on Systems, Man, and Cybernetics: Systems*, 47(8): 2301–2310, 2017.

94 Dongsheng Guo, Xinjie Lin, Zhaozhu Su, Sibo Sun, and Zhijing Huang. Design and analysis of two discrete-time ZD algorithms for time-varying nonlinear minimization. *Numerical Algorithms*, 77: 23–36, 2018.

95 Yunong Zhang, Huihui Gong, Min Yang, Jian Li, and Xuyun Yang. Stepsize range and optimal value for Taylor–Zhang discretization formula applied to zeroing neurodynamics illustrated via future equality-constrained quadratic programming. *IEEE Transactions on Neural Networks and Learning Systems*, 30(3): 959–966, 2018.

96 Hari Teja Kalidindi, Thomas George Thuruthel, Cecilia Laschi, and Egidio Falotico. Cerebellum-inspired approach for adaptive kinematic control of soft robots. In *IEEE International Conference on Soft Robotics*, IEEE, pages 684–689, 2019.

97 Xin Luo, Ye Yuan, MengChu Zhou, Zhigang Liu, and Mingsheng Shang. Non-negative latent factor model based on β-divergence for recommender systems. *IEEE Transactions on Systems, Man, and Cybernetics: Systems*, 51(8): 4612–4623, 2021.

98 Shangke Lyu and Chien Chern Cheah. Data-driven learning for robot control with unknown Jacobian. *Automatica*, 120: 109120, 2020.

99 Zhijun Zhang, Zhijun Li, Yunong Zhang, Yamei Luo, and Yuanqing Li. Neural-dynamic-method-based dual-arm CMG scheme with time-varying constraints applied to humanoid robots. *IEEE Transactions on Neural Networks and Learning Systems*, 26(12): 3251–3262, 2015.

100 Dechao Chen and Yunong Zhang. Robust zeroing neural-dynamics and its time-varying disturbances suppression model applied to mobile robot manipulators. *IEEE Transactions on Neural Networks and Learning Systems*, 29(9): 4385–4397, 2018.

101 Zhengtai Xie, Long Jin, Xin Luo, Bin Hu, and Shuai Li. An acceleration-level data-driven repetitive motion planning scheme for kinematic control of robots with unknown structure. *IEEE Transactions on Systems, Man, and Cybernetics: Systems*, 52(9): 5679–5691, 2021.

102 Zhijia Zhao and Zhijie Liu. Finite-time convergence disturbance rejection control for a flexible Timoshenko manipulator. *IEEE/CAA Journal of Automatica Sinica*, 8(1): 157–168, 2021.

103 Guangdeng Zong, Hangli Ren, and Hamid Reza Karimi. Event-triggered communication and annular finite-time H_∞ filtering for networked switched systems. *IEEE Transactions on Cybernetics*, 51(1): 309–317, 2021.

104 Long Jin, Yunong Zhang, and Shuai Li. Integration-enhanced Zhang neural network for real-time-varying matrix inversion in the presence of various kinds of noises. *IEEE Transactions on Neural Networks and Learning Systems*, 27(12): 2615–2627, 2016.

105 Ning Tan, Peng Yu, Zhiyan Zhong, and Fenglei Ni. A new noise-tolerant dual-neural-network scheme for robust kinematic control of robotic arms with unknown models. *IEEE/CAA Journal of Automatica Sinica*, 9(10): 1778–1791, 2022.

106 Ning Tan, Peng Yu, Xinyu Zhang, and Tao Wang. Model-free motion control of continuum robots based on a zeroing neurodynamic approach. *Neural Networks*, 133: 21–31, 2021. ISSN 18792782.

107 Muhammad Sunny Nazeer, Cecilia Laschi, and Egidio Falotico. RL-based adaptive controller for high precision reaching in a soft robot arm. *IEEE Transactions on Robotics*, 40: 2498–2512, 2024. https://doi.org/10.1109/TRO.2024.3381558.

108 Xin Ma, Xuchen Wang, Zihao Zhang, Puchen Zhu, Shing Shin Cheng, and Kwok Wai Samuel Au. Design and experimental validation of a novel hybrid continuum robot with enhanced dexterity and manipulability in confined space. *IEEE/ASME Transactions on Mechatronics*, 28(4): 1826–1835, 2023. https://doi.org/10.1109/TMECH.2023.3276076.

109 Shotaro Shimegi, Keitaro Ishibashi, and Hiroyuki Ishii. Development of wearable lightweight continuum manipulator to enhance capability of human arm. In *IEEE International Conference on Mechatronics and Automation*, IEEE, pages 337–344, 2024.

110 Tommaso Ranzani, Giada Gerboni, Matteo Cianchetti, and Arianna Menciassi. A bioinspired soft manipulator for minimally invasive surgery. *Bioinspiration & Biomimetics*, 10(3): 035008, 2015.

111 Hangjie Mo, Ruofeng Wei, Xiaowen Kong, Yujia Zhai, Yunhui Liu, and Dong Sun. Data-efficient learning control of continuum robots in constrained environments. *IEEE Transactions on Automation Science and Engineering*, 22: 984–995, 2024. https://doi.org/10.1109/TASE.2024.3357816.

112 Ke Qiu, Hongye Zhang, Jingyu Zhang, Rong Xiong, Haojian Lu, and Yue Wang. An actuator space optimal kinematic path tracking framework for tendon-driven continuum robots: Theory, algorithm and validation. *The International Journal of Robotics Research*, 44(6): 1006–1034, 2025. https://doi.org/10.1177/02783649241290525.

113 Balint Thamo, Kev Dhaliwal, and Mohsen Khadem. Rapid solution of cosserat rod equations via a nonlinear partial observer. In *IEEE International Conference on Robotics and Automation*, pages 9433–9438, 2021.

114 Changyeob Baek, Kyungho Yoon, and Do-Nyun Kim. Finite element modeling of concentric-tube continuum robots. *Structural Engineering and Mechanics*, 57(5): 809–821, 2016.

115 Bingbin Yu, Joseỉ de Gea Fernaỉndez, and Tao Tan. Probabilistic kinematic model of a robotic catheter for 3d position control. *Soft Robotics*, 6(2): 184–194, 2019.

116 Jaewoong Jung, Myungsun Park, DongWook Kim, and Yong-Lae Park. Optically sensorized elastomer air chamber for proprioceptive sensing of soft pneumatic actuators. *IEEE Robotics and Automation Letters*, 5(2): 2333–2340, 2020.

117 Michele Giorelli, Federico Renda, Marcello Calisti, Andrea Arienti, Gabriele Ferri, and Cecilia Laschi. Neural network and Jacobian method for solving the inverse statics of a cable-driven soft arm with nonconstant curvature. *IEEE Transactions on Robotics*, 31(4): 823–834, 2015.

118 Thomas George Thuruthel, Benjamin Shih, Cecilia Laschi, and Michael Thomas Tolley. Soft robot perception using embedded soft sensors and recurrent neural networks. *Science Robotics*, 4(26): eaav1488, 2019.

119 Hao Jiang, Zhanchi Wang, Yusong Jin, Xiaotong Chen, Peijin Li, Yinghao Gan, Sen Lin, and Xiaoping Chen. Hierarchical control of soft manipulators towards unstructured interactions. *The International Journal of Robotics Research*, 40(1): 411–434, 2021.

120 Yunong Zhang, Danchi Jiang, and Jun Wang. A recurrent neural network for solving Sylvester equation with time-varying coefficients. *IEEE Transactions on Neural Networks*, 13(5): 1053–1063, 2002.

121 Dechao Chen, Yunong Zhang, Shuai Li, and Yihong Ling. Robust Zhang neural network for tracking control of parallel robot manipulators with unknown parameters. In *Chinese Control And Decision Conference*, pages 3527–3532, 2019. ISBN 9781728101057.

122 Zhijun Zhang and Ziyi Yan. A varying parameter recurrent neural network for solving nonrepetitive motion problems of redundant robot manipulators. *IEEE Transactions on Control Systems Technology*, 23(2): 679–689, 2018.

123 Xiaosong Hu, Fengchun Sun, Yuan Zou, and Huei Peng. Online estimation of an electric vehicle Lithium-Ion battery using recursive least squares with forgetting. In *Proceedings of the 2011 American Control Conference*, pages 935–940, 2011.

124 Run-Ning Zhang, Peng Yu, and Ning Tan. Continuum robot control based on varying parameter recursive network and recursive least square. *Acta Automatica Sinica*, 51(1): 90–103, 2025. https://doi.org/10.16383/j.aas.c240469.

125 Bolin Liao and Yunong Zhang. Different complex ZFs leading to different complex ZNN models for time-varying complex generalized inverse matrices. *IEEE Transactions on Neural Networks and Learning Systems*, 25(9): 1621–1631, 2013.

126 Ning Tan, Peng Yu, and Fenglei Ni. New varying-parameter recursive neural networks for model-free kinematic control of redundant manipulators with limited measurements. *IEEE Transactions on Instrumentation and Measurement*, 71: 1–14, 2022.

127 Lin Xiao, Xiaopeng Li, Pengling Cao, Yongjun He, Wensheng Tang, Jichun Li, and Yaonan Wang. A dynamic-varying parameter enhanced ZNN model for solving time-varying complex-valued tensor inversion with its application to image encryption. *IEEE Transactions on Neural Networks and Learning Systems*, 35(10): 13681–13690, 2024.

128 Chun Yuan Deng. A generalization of the Sherman–Morrison–Woodbury formula. *Applied Mathematics Letters*, 24(9): 1561–1564, 2011.

129 Pu Yang and Yuxin Su. Proximate fixed-time prescribed performance tracking control of uncertain robot manipulators. *IEEE/ASME Transactions on Mechatronics*, 27(5): 3275–3285, 2022.

130 Jiangtao Han, Zhiji Han, and Zhijie Liu. Adaptive control for a constrained soft manipulator with prescribed performance. *IFAC-PapersOnLine*, 53(5): 524–529, 2020.

131 Yan Yang, Jiangtao Han, Zhijie Liu, Zhijia Zhao, and Keum-Shik Hong. Modeling and adaptive neural network control for a soft robotic arm with prescribed motion constraints. *IEEE/CAA Journal of Automatica Sinica*, 10(2): 501–511, 2023.

132 Yisen Huang, Jian Li, Xue Zhang, Ke Xie, Jixiu Li, Yue Liu, Calvin Sze Hang Ng, Philip Wai Yan Chiu, and Zheng Li. A surgeon preference-guided autonomous instrument tracking method with a robotic flexible endoscope based on dVRK platform. *IEEE Robotics and Automation Letters*, 7(2): 2250–2257, 2022.

133 Konrad Leibrandt, Christos Bergeles, and Guang-Zhong Yang. Concentric tube robots: Rapid, stable path-planning and guidance for surgical use. *IEEE Robotics & Automation Magazine*, 24(2): 42–53, 2017.

134 Sven Lilge, Timothy David Barfoot, and Jessica Burgner-Kahrs. Continuum robot state estimation using Gaussian process regression on SE(3). *The International Journal of Robotics Research*, 41(13-14): 1099–1120, 2022.

135 Weibing Li. Predefined-time convergent neural solution to cyclical motion planning of redundant robots under physical constraints. *IEEE Transactions on Industrial Electronics*, 67(12): 10732–10743, 2020.

136 Li Dongyu, Shuzhi Sam Ge, and Tong Heng Lee. *Time-Synchronized Control: Analysis and Design*. Springer, 2022.

137 Ning Tan, Peng Yu, Xin Wang, and Kai Huang. Behavior-predefined adaptive control for heterogeneous continuum robots. *The International Journal of Robotics Research*, 44(1): 65–95, 2025.

138 Ning Tan, Peng Yu, and Kai Huang. Predefined-time convergent motion control for heterogeneous continuum robots. In *Robotics: Science and Systems*, Daegu, Republic of Korea, 2023.

139 Ryuichi Ashino, Michihiro Nagase, and Rémi Vaillancourt. Behind and beyond the MATLAB ODE suite. *Computers & Mathematics with Applications*, 40(4-5): 491–512, 2000.

140 Dongyu Li, Haoyong Yu, Keng Peng Tee, Yan Wu, Shuzhi Sam Ge, and Tong Heng Lee. On time-synchronized stability and control. *IEEE Transactions on Systems, Man, and Cybernetics: Systems*, 52(4): 2450–2463, 2022.

141 Peng Yu, Siming Ma, and Ning Tan. Predefined-time-synchronized whole-body control of velocity-controlled mobile manipulators with unknown kinematics. *IEEE Transactions on Industrial Electronics*, 2025. https://doi.org/10.1109/TIE.2025.3553164.

142 Charles Van Loan. The ubiquitous Kronecker product. *Journal of Computational and Applied Mathematics*, 123(1-2): 85–100, 2000.

143 Zhijun Zhang and Ziyi Yan. An adaptive fuzzy recurrent neural network for solving the nonrepetitive motion problem of redundant robot manipulators. *IEEE Transactions on Fuzzy Systems*, 28: 684–691, 2020.

144 Lei Jia, Lin Xiao, Jianhua Dai, Zhao Qi, Zhijun Zhang, and Yongsheng Zhang. Design and application of an adaptive fuzzy control strategy to zeroing neural network for solving time-variant QP problem. *IEEE Transactions on Fuzzy Systems*, 29: 1544–1555, 2021.

145 Peng Qi, Chuang Liu, Ahmad Ataka, Hak-Keung Lam, and Kaspar Althoefer. Kinematic control of continuum manipulators using a fuzzy-model-based approach. *IEEE Transactions on Industrial Electronics*, 63(8): 5022–5035, 2016.

146 Farzin Piltan,Cheol-Hong Kim, and Jong-Myon Kim. Adaptive fuzzy-based fault-tolerant control of a continuum robotic system for maxillary sinus surgery. *Applied Sciences*, 9(12): 2490, 2019.

147 Ning Tan, Zixiao Ye, Peng Yu, and Fenglei Ni. A dual fuzzy-enhanced neurodynamic scheme for model-less kinematic control of redundant and hyperredundant robots. *IEEE Transactions on Fuzzy Systems*, 30(10): 4409–4422, 2022. https://doi.org/10.1109/TFUZZ.2022.3152077.

148 Kaushik Roy, Akhilesh Jaiswal, and Priyadarshini Panda. Towards spike-based machine intelligence with neuromorphic computing. *Nature*, 575(7784): 607–617, 2019.

149 Wolfgang Maass. Networks of spiking neurons: The third generation of neural network models. *Neural Networks*, 10(9): 1659–1671, 1997. ISSN 0893-6080.

150 Wolfgang Maass, Thomas Natschläger, and Henry Markram. Real-time computing without stable states: A new framework for neural computation based on perturbations. *Neural Computation*, 14(11): 2531–2560, 2002.

151 Laurent Ott, Florent Nageotte, Philippe Zanne, and Michel de Mathelin. Robotic assistance to flexible endoscopy by physiological-motion tracking. *IEEE Transactions on Robotics*, 27(2): 346–359, 2011.

152 Haoyuan Gu, Hesheng Wang, Fan Xu, Zhe Liu, and Weidong Chen. Active fault detection of soft manipulator in visual servoing. *IEEE Transactions on Industrial Electronics*, 68(10): 9778–9788, 2021.

153 Juan Sandoval, Hang Su, Pierre Vieyres, Gérard Poisson, Giancarlo Ferrigno, and Elena De Momi. Collaborative framework for robot-assisted minimally invasive surgery using a 7-DoF anthropomorphic robot. *Robotics and Autonomous Systems*, 106: 95–106, 2018. ISSN 0921-8890.

154 Weibing Li, Philip Wai Yan Chiu, and Zheng Li. An accelerated finite-time convergent neural network for visual servoing of a flexible surgical endoscope with physical and RCM constraints. *IEEE Transactions on Neural Networks and Learning Systems*, 31(12): 5272–5284, 2020.

155 Hamid Sadeghian, Fatemeh Zokaei, and Shahram Hadian Jazi. Constrained kinematic control in minimally invasive robotic surgery subject to remote center of motion constraint. *Journal of Intelligent & Robotic Systems*, 95(3): 901–913, 2019. ISSN 1573-0409.

156 Hang Su, Juan Sandoval, Pierre Vieyres, Gérard Poisson, Giancarlo Ferrigno, and Elena D. Momi. Safety-enhanced collaborative framework for tele-operated minimally invasive surgery using a 7-DoF torque-controlled robot. *International Journal of Control, Automation and Systems*, 16(6): 2915–2923, 2018.

157 Yongqing Fan, Yue An, Wenqing Wang, and Chenguang Yang. T-S fuzzy adaptive control based on small gain approach for an uncertain robot manipulators. *International Journal of Fuzzy Systems*, 22(3): 930–942, 2020. ISSN 2199-3211.

158 Ning Tan and Peng Yu. Robust model-free control for redundant robotic manipulators based on zeroing neural networks activated by nonlinear functions. *Neurocomputing*, 438: 44–54, 2021.

159 Xin Wang and Ning Tan. Cerebellum-inspired model-free tracking control and visual servoing of a rigid-flexible hybrid robotic endoscope with RCM constraints. *IEEE Transactions on Industrial Electronics*, 70(12): 12626–12635, 2023. https://doi.org/10.1109/TIE.2023.3239875.

160 Xingting Yao, Fanrong Li, Zitao Mo, and Jian Cheng. Glif: A unified gated leaky integrate-and-fire neuron for spiking neural networks. *Advances in Neural Information Processing Systems*, 35: 32160–32171, 2022.

161 Weibing Li, Chengzhi Song, and Zheng Li. An accelerated recurrent neural network for visual servo control of a robotic flexible endoscope with joint limit constraint. *IEEE Transactions on Industrial Electronics*, 67(12): 10787–10797, 2020.

162 Bruno Siciliano and Jean-Jacques Slotine. A general framework for managing multiple tasks in highly redundant robotic systems. In *Fifth International Conference on Advanced Robotics \'Robots in Unstructured Environments, volume 2*, pages 1211–1216, 1991.

163 Yiping Xie, Xilong Hou, and Shuangyi Wang. Design of a novel haptic joystick for the teleoperation of continuum-mechanism-based medical robots. *Robotics*, 12(2): 52, 2023.

164 Dengliang Lin, Chenguang Yang, and Shilu Dai. Teleoperation control of flexible joint continuum robot. In *International Conference on Control Science and Systems Engineering*, pages 57–62, 2021.

165 Long Jin, Jiazheng Zhang, Xin Luo, Mei Liu, Shuai Li, Lin Xiao, and Zihao Yang. Perturbed manipulability optimization in a distributed network of redundant robots. *IEEE Transactions on Industrial Electronics*, 68(8): 7209–7220, 2021.

166 Zhijun Zhang, Tingzhong Fu, Ziyi Yan, Long Jin, Lin Xiao, Yuping Sun, Zhuliang Yu, and Yuanqing Li. A varying-parameter convergent-differential neural network for solving joint-angular-drift problems of redundant robot manipulators. *IEEE/ASME Transactions on Mechatronics*, 23(2): 679–689, 2018.

167 Emanuel SlawiNski and Vicente Mut. Transparency in time for teleoperation systems. In *International Conference on Robotics and Automation*, pages 200–205, 2008.

168 Phongsaen Pitakwatchara. Control of time-varying delayed teleoperation system using corrective wave variables. In *IEEE/RSJ International Conference on Intelligent Robots and Systems*, pages 4550–4555, 2015.

169 Ning Tan, Peng Yu, Mao Zhang, and Changsheng Li. Toward unified adaptive teleoperation based on damping ZNN for robot manipulators with unknown kinematics. *IEEE Transactions on Industrial Electronics*, 70(9): 9227–9236, 2023. https://doi.org/10.1109/TIE.2022.3210522.

Index

Printed and bound by CPI Group (UK) Ltd, Croydon, CR0 4YY

17/09/2026

14964235-0002